Nanomagnetic Actuation in Biomedicine

Nanomagnetic Actuation in Biomedicine
Basic Principles and Applications

Edited by
Jon Dobson and Carlos Rinaldi

CRC Press
Taylor & Francis Group
Boca Raton London New York

CRC Press is an imprint of the
Taylor & Francis Group, an **informa** business

CRC Press
Taylor & Francis Group
6000 Broken Sound Parkway NW, Suite 300
Boca Raton, FL 33487-2742

First issued in paperback 2019

Library of Congress Cataloging-in-Publication Data

Names: Dobson, Jon, editor. | Rinaldi, Carlos (Chemical engineer), editor.
Title: Nanomagnetic actuation in biomedicine : basic principles and   applications / [edited by] Jon Dobson and Carlos Rinaldi.
Description: Boca Raton : Taylor & Francis, 2017. | Includes bibliographical references and index.
Identifiers: LCCN 2017030045| ISBN 9781466591219 (hardback : alk. paper) | ISBN 9781315374086 (ebook)
Subjects: | MESH: Receptors, Cell Surface–metabolism | Magnetite Nanoparticles | Signal Transduction–physiology | Theranostic Nanomedicine–methods
Classification: LCC QP517.C45 | NLM QU 55.7 | DDC 571.7/4–dc23
LC record available at https://lccn.loc.gov/2017030045

Visit the Taylor & Francis Web site at
http://www.taylorandfrancis.com

and the CRC Press Web site at
http://www.crcpress.com

Contents

Editors vii

Contributors ix

1. Nanomagnetic Actuation in Biomedicine: An Introduction 1
   Adam Monsalve, Carlos Rinaldi, and Jon Dobson

2. Synthesis and Surface Functionalization of Ferrite Nanoparticles 9
   Jennifer S. Andrew, Carlos Rinaldi, and O. Thompson Mefford

3. Thermomagnetic Activation of Cellular Ion Channels to
   Control Cellular Activity 41
   Heng Huang and Arnd Pralle

4. Magnetic Twisting Cytometry 59
   Daniel Isabey, Christelle Angely, Adam Caluch,
   Bruno Louis, and Gabriel Pelle

5. Magnetic Needle Development 93
   David Arnold, Zak Kaufman, and Alexandra Garraud

6. Magnetic Delivery of Cell-Based Therapies 123
   Boris Polyak and Richard Sensenig

7. Magnetic Capture and Actuation of Thermosensitive Drug
   Carriers Using Iron Oxide Nanoparticles 147
   Mary Kathryn Sewell-Loftin, Mary L. Hampel, Amy E. Frees,
   Lauren M. Blue, Natalie Lapp, Minghua Zhang, Jaimee M. Robertson,
   Rhythm R. Shah, and Christopher S. Brazel

8. Magnetic Cell Patterning 175
   Thomas Crawford

9. Applications of Magnetic Nanoparticles in Tissue Engineering
   and Regenerative Medicine 205
   James R. Henstock, Hareklea Markides, Hu Bin,
   Alicia J. El Haj, and Jon Dobson

10. Magnetic Nanoparticles for 3D Cell Culture 229
    Hubert Tseng, Robert M. Raphael, Thomas C. Killian, and Glauco R. Souza

Index 245

Contents

Editors

Jon Dobson graduated with a BSc and MSc in Geology and Geophysics from the University of Florida. He obtained his PhD in Natural Sciences in 1991 from the Swiss Federal Institute of Technology, ETH-Zurich. He did his post-doctoral training in geophysics and biophysics at both the ETH-Zurich and The University of Western Australia, before taking a faculty position at Keele University in the United Kingdom. In 2011, he returned to the University of Florida as Professor of Biomaterials and Biomedical Engineering, and founding director of the Institute for Cell and Tissue Science and Engineering (ICTSE) at UF.

Dr. Dobson's research focuses on biomedical applications of magnetic micro- and nanoparticles, the role of brain iron in neurodegenerative diseases, and biomedical device design. He is a Fellow of the American Association for the Advancement of Science (AAAS), the American Institute for Medical and Biological Engineering (AIMBE), The Royal Society of Biology, The Royal Society of Medicine, and a past Royal Society of London Wolfson Research Merit Fellow. In 2002, he was selected for the Wellcome Trust's *Sir Henry Wellcome Showcase Award*, and in 2008 the UK Medical Research Council's *César Milstein Award*. He has authored or co-authored more than 190 peer-reviewed publications, including the 2nd most highly cited paper in the field, has 20 patents awarded or pending, and is co-founder of three spin-off companies.

Carlos Rinaldi received his BS degree in Chemical Engineering from the University of Puerto Rico, Mayagüez (1998), and completed degrees in Master of Science in Chemical Engineering (2001), Master of Science in Chemical Engineering Practice (2001), and Doctor of Philosophy (2002) in Chemical Engineering at the Massachusetts Institute of Technology (MIT). During the summer of 2002 he served as Assistant Station Director for the David H. Koch School of Chemical Engineering Practice at MIT. From 2002 to 2012 he was a professor in the Department of Chemical Engineering at the University of Puerto Rico, Mayagüez. In 2012 he joined the University of Florida, with joint appointments in the J. Crayton Pruitt Family Department of Biomedical Engineering and the Department of Chemical Engineering. Carlos Rinaldi has authored and co-authored more than 120 publications in the fields of magnetic nanomaterials, fluid mechanics of magnetic nanoparticle suspensions, and nanotechnology. His research interests are in biomedical applications of magnetic nanoparticles and the fluid physics of magnetic nanoparticle suspensions. He received the Presidential Early Career Award for Scientists and Engineers (PECASE) in 2005, in recognition of his contributions to magnetic nanoparticle research and to broadening participation of under-represented groups in engineering.

Contributors

Jennifer S. Andrew
Departments of Biomedical
  Engineering and Chemical
  Engineering
University of Florida
Gainesville, Florida

Christelle Angely
Department of Cell and Respiratory
  Mechanics
Université Paris Est
Créteil Cedex, France

David Arnold
Department of Electrical and
  Computer Engineering
University of Florida
Gainesville, Florida

Hu Bin
National Centre for Nanoscience
  and Technology
Beijing, China

Lauren M. Blue
Department of Chemical and
  Biological Engineering
University of Alabama
Tuscaloosa, Alabama

Christopher S. Brazel
Department of Chemical and
  Biological Engineering
University of Alabama
Tuscaloosa, Alabama

Adam Caluch
Department of Cell and Respiratory
  Mechanics
Université Paris Est
Créteil Cedex, France

Thomas Crawford
Department of Physics
University of South Carolina
Columbia, South Carolina

Jon Dobson
University of Florida
J Crayton Pruitt Family Professor of
  Biomedical Engineering
Departments of Biomedical
  Engineering and Materials
  Science & Engineering
Gainesville, Florida

Alicia J. El Haj
Institute for Science and Technology
  in Medicine
Keele University
Keele, England

Amy E. Frees
Department of Chemical and
  Biological Engineering
University of Alabama
Tuscaloosa, Alabama

Alexandra Garraud
Department of Electrical and
  Computer Engineering
University of Florida
Gainesville, Florida 32611
USA

Mary L. Hampel
Department of Chemical and
    Biological Engineering
University of Alabama
Tuscaloosa, Alabama

James R. Henstock
Institute for Ageing and Chronic
    Disease
University of Liverpool
Liverpool, England

Heng Huang
Department of Physics, Biophysics,
    and Physiology
University of Buffalo
Buffalo, New York

Daniel Isabey
Department of Cell and Respiratory
    Mechanics
Université Paris Est
Créteil Cedex, France

Zak Kaufman
Department of Electrical and
    Computer Engineering
University of Florida
Gainesville, Florida

Thomas C. Killian
Department of Physics
Rice University
Houston, Texas

Natalie Lapp
Department of Chemical and
    Biological Engineering
University of Alabama
Tuscaloosa, Alabama

Bruno Louis
Department of Cell and Respiratory
    Mechanics
Université Paris Est
Créteil Cedex, France

Hareklea Markides
Institute for Science and Technology
    in Medicine
Keele University
Keele, England

O. Thompson Mefford
Department of Materials Science
    and Engineering
Clemson University
Clemson, South Carolina

Adam Monsalve
J. Crayton Pruitt Family Department
    of Biomedical Engineering
University of Florida
Gainesville, Florida

Gabriel Pelle
Department of Cell and Respiratory
    Mechanics
AP–HP, Groupe Hospitalier
    H. Mondor—A. Chenevier, CHI
    Créteil, Service de Physiologie—
    Explorations Fonctionnelles
Université Paris Est
Créteil Cedex, France

Boris Polyak
Department of Surgery and
    Department of Pharmacology
    and Physiology
Drexel University College of
    Medicine
Philadelphia, Pennsylvania

Arnd Pralle
Department of Physics, Biophysics,
and Physiology
University of Buffalo
Buffalo, New York

Robert M. Raphael
Department of Bioengineering
Rice University
Houston, Texas

Carlos Rinaldi
J. Crayton Pruitt Family Department
of Biomedical Engineering
Department of Chemical
Engineering
University of Florida
Gainesville, Florida

Jaimee M. Robertson
Department of Chemical and
Biological Engineering
University of Alabama
Tuscaloosa, Alabama

Richard Sensenig
Department of Surgery
University of Pennsylvania
Philadelphia, Pennsylvania

Mary Kathryn Sewell-Loftin
Department of Chemical and
Biological Engineering
University of Alabama
Tuscaloosa, Alabama

Rhythm R. Shah
Department of Chemical and
Biological Engineering
University of Alabama
Tuscaloosa, Alabama

Glauco R. Souza
Nano 3D Biosciences
Houston, Texas and
Departments of Bioengineering and
Physics
Rice University
Houston, Texas

Hubert Tseng
Nano 3D Biosciences
Houston, Texas and
Departments of Bioengineering and
Physics
Rice University
Houston, Texas

Minghua Zhang
Department of Chemical and
Biological Engineering
University of Alabama
Tuscaloosa, Alabama

1

Nanomagnetic Actuation in Biomedicine: An Introduction

Adam Monsalve, Carlos Rinaldi, and Jon Dobson

CONTENTS

1.1 Magnetism and Biomedicine 1
1.2 Micro- and Nanoscale Magnetic Bio-Actuators 4
References..6

1.1 Magnetism and Biomedicine

The human body is largely transparent to magnetic fields. This is not the case for electric fields, as anyone who has touched a live wire can attest. That does not, however, mean that the human body has no interaction with strong magnetic fields. This unique ability of magnetic fields to penetrate into or pass through the body, coupled with our ability to understand the interactions of these fields with the materials that make up our bodies, has enabled the development of such transformative technologies as magnetic resonance imaging (MRI).

Strong, static magnetic fields have the ability to force into alignment, to a degree, precessing hydrogen protons (which are spinning like tops) in the water molecules within our body. By briefly adding an alternating frequency (AC) magnetic field at a frequency that matches the natural precession frequency of these protons in the body's water molecules (the Larmor precession frequency)—and by orienting that AC field perpendicular to the static field—it is possible to couple the AC field to these precessing protons and knock them out of alignment with the static field and into the plane of the AC field. This is akin to pushing a child on a swing. If you stand back and push the child when she is at the apex of her arc, it is easy to transfer energy to her and keep her swinging. However, if you stand directly below the crossbar to which the swing is attached and try to stop the rearward motion and reverse the child's direction, much more energy is required—in fact, she may knock you over. This basically describes the principle of "resonance"—a natural frequency at which something vibrates or oscillates.

When the AC field is removed, the proton spins are free to spin back into alignment with the static field, giving off a radio signal that is indicative of the water molecule's environment and, with the help of magnetic field gradients and some sophisticated algorithms, an image of structures within the body. In this case, the applied magnetic fields—both the static and the AC fields—have interacted with the protons without touching them. This "action at a distance" is what makes magnetism and magnetic materials uniquely powerful for the development of tools for biomedicine.

We can take this ability of a magnetic field to remotely interact with materials a bit further. If we introduce magnetic material (for example, micro- or nanoparticles and structures) into the body, we can essentially reach into the body with magnetic fields and actuate (move) or deposit energy into these particles. Furthermore, if the particles are coated with biocompatible polymers to which biomolecular recognition sequences can be attached, it is possible to target these particles to specific receptors on the surface of the cells. Once attached, the externally applied fields can be used to transfer energy to these receptors, activating biochemical signaling cascades within the cells. The activation of these signaling cascades controls cell functions, such as cell division, cell–cell communication, motility, tissue matrix deposition, and apoptosis, among others. It is also possible to move or target the cells to a specific location (perhaps a site of injury) in the body. Essentially, these magnetic particles can be thought of as "remote control" switches for cellular functions. It is even possible to move and pattern cells for potential applications in tissue engineering using these principles.

In addition, by coupling these magnetic particles within the body to radio-frequency (RF) fields, it is possible to deposit heat into a tumor to kill it (if you have enough material within the tumor) via necrosis, lysosomal death pathways, or activation of apoptosis signaling. Currently, the mechanisms behind this technique—magnetic fluid hyperthermia—are somewhat controversial as energy transfer to specific cell-surface receptors has been shown to activate apoptosis without any measureable macroscopic heating.[1,2] However, this too can be a powerful tool for controlling cell behavior as no high field gradients are required to pull on the particle actuators.

The physical principles of these processes are discussed in more detail in numerous publications and in later chapters of this book (e.g., Pankhurst et al.[3,4]). For now, we take a step back and have a brief look at magnetism and magnetic materials and a few of their potential biomedical applications that exploit remote interactions between fields and materials.

While MRI and interactions of strong magnetic fields with proton spins was used as an example earlier, this book focuses primarily on the interactions of magnetic fields with "ferromagnetic" materials—materials whose magnetic properties are dominated by unpaired electron spins. Ferromagnetic materials have large magnetic moments resulting from coupling of many unpaired electron spins within the material or particle. There are several subclasses of ferromagnetism and we primarily focus on materials that are either antiferromagnetic (electron spins are coupled but oriented antiparallel in neighboring lattices) or ferromagnetic (spins are antiparallel but there is an unequal number, giving rise to a strong magnetization in the direction of the lattices with an extra spin, or Bohr magneton) (Figure 1.1).

In order to use magnetic particles in the body, the specific application must be considered very carefully. If the particles are to be injected intravenously, there is a risk of aggregation of the particles due to magnetic dipole interactions. This can be detrimental to the subject and in extreme cases could result in the formation of an embolism. For this reason, superparamagnetic particles are primarily used for in vivo biomedical applications. Superparamagnetic particles have strong magnetization (many unpaired electron spins that are coupled), however, in the absence of a field, they are essentially in a zero magnetic remanence ground state as the energy associated with the exchange-coupled electron spins is lower than thermal energy in the system under consideration (Figure 1.2). This phenomenon is governed by the Neél-Arrhenius relationship:

$$\tau = \tau_0 \exp(KV/K_B T)$$

where τ is the magnetic relaxation time (the time that the particle spends "magnetized" in a specific direction), τ_0 is the attempt time (a material constant), K is the magnetic anisotropy constant for the material, V is the volume, K_B is Boltzmann's constant, and T is the temperature. (For a more

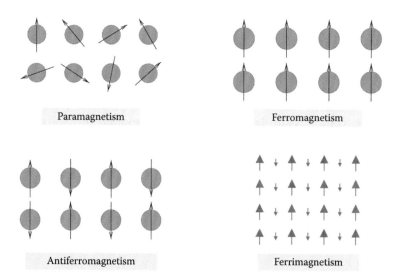

FIGURE 1.1
The magnetic field generated by the spin of an electron is known as the Bohr magneton—a fundamental magnetic unit that gives rise to magnetization in materials. This figure shows the electronic spin structure of four classes of materials with uncompensated electron spins. In paramagnets, these spins are not coupled, while in the ferro-, antiferro-, and ferri-magnetic materials, the spins are exchange coupled, leading to stronger magnetization and magnetic remanence due to the addition of many Bohr magnetons with the same orientation.

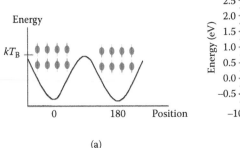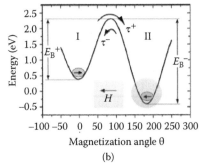

FIGURE 1.2

In superparamagnetic materials, although the electron spins are coupled, thermal energy is greater than magnetic anisotropy energy, leading to (a) "flipping" of the group of spins back and forth along the easy axis of magnetization. (b) Once a magnetic field is applied, the asymmetry of the energy "landscape" leads to alignment of the spins with the applied field. Once the field is removed, the magnetization is lost.

thorough discussion of magnetic materials, including superparamagnetism, the reader is referred to textbooks by Jiles[5] and Craik.)[6]

This means that they don't behave like larger, stable magnets and have less tendency to aggregate. For example, ferromagnetic T_2 MRI contrast agents are generally superparamagnetic iron oxides complexed with a sugar such as dextran.

Iron oxides are used in many in vivo applications as they are thought to be largely biocompatible—at least more biocompatible than many other strongly magnetic materials such as FePt or cobalt ferrites. Of the iron oxides available, magnetite (Fe_3O_4) and its oxidation product, maghemite (γFe_2O_3) are most commonly used. These are both ferrimagnetic materials and their magnetization is similar; ~90 emu/g for magnetite and ~80 emu/g for maghemite. For these materials, the superparamagnetic size limit is around 30 nm (depending on shape as they have, strong shape anisotropy can act to stabilize the magnetization even in the absence of an applied field). For these reasons, with some notable exceptions that we discuss, nanoscale magnetic particles and structures are used for the biomedical actuation applications explored in this book. The synthesis and functionalization of such particles are discussed in detail in Chapter 2.

1.2 Micro- and Nanoscale Magnetic Bio-Actuators

Though this book focuses primarily on using magnetic fields to deliver energy to particles to either actuate cellular processes, target cells or molecules to sites within the body, or trigger release of therapeutics, magnetic microparticles were first used to investigate biological systems as early as the 1920s. In 1922, Heilbrunn introduced iron microparticles into slime molds and pulled on them with high-gradient magnetic fields. By measuring the speed of the particles moving along the field gradient, he was able to investigate the viscosity of the slime mold "protoplasm."[7] A couple of years later (1924), William Seifritz, built a microdissection device and conducted experiments on cells based on Heilbrunn's earlier results.[8] He was able to inject large (9–15 µm) nickel particles into cells and, using an electromagnet mounted on the microscope, made measurements of the rheological properties of cells. The results varied significantly between experiments and were at odds with other methods, so the technique did not gain wide acceptance.

After languishing for 25 years, the use of magnetic particles to investigate cells got another kick-start from work by Francis Crick and AFW Hughes. Rather than injecting particles into cells, they incubated chick fibroblasts with 2–10 µm magnetite particles, allowing the cells to phagocytose (internalize) them. Once internalized, they examined twisting, dragging, and prodding motions of the particles under the influence of an applied field to again measure rheological properties of the cytoplasm.[9]

In the 1980s and 1990s, after another decade-long break in this research, Valberg, Butler, Wang, Ingber, and others began to bind magnetic microparticles to cell-surface receptors—mainly integrins—in order to investigate the mechanical properties of the cytoskeleton to which the integrins are connected.[10–15] The work of Valberg and Butler developed into the technique of "magnetic twisting cytometry" (MTC), which is explored in more detail in Chapter 4.

In the later 1990s, magnetic micro- and magnetic nanoparticle (MNP)-based activation of cell signaling began to take off. Several groups, led by Pommerenke, Sackmann, Glogauer, Ferrier, and others, combined the ideas of attaching particles to surface receptors (the technique that started with MTC) and, rather than twisting the particles, applied a translational force to the particles by exposing them to high-gradient magnetic fields. This force on the particles acts to mechanically deform the cell membrane, activating nearby mechanosensitive ion channels present on the surface of the cells.[16–20]

This work would eventually lead to work by Dobson and collaborators on targeting and remotely activating specific cell-surface receptors using MNPs—a technique termed Magnetic Activation of Receptor Signaling (MARS) or Magnetic Ion Channel Activation (MICA).[21–23] That work is explored in more detail in Chapter 9. In order to advance this technique, novel magnetic actuation systems are being developed that now enable precise field focusing for *in vitro* studies, as is described in Chapter 5, as well as novel magnet arrays and the use of RF fields to deposit energy into MNPs coupled to cell-surface receptors to control whole organisms (Chapter 3).

More recently, the biomedical uses of magnetic micro- and MNPs has expanded to include such novel techniques as magnetically activated drug and biomolecule release, which is described in Chapter 7, and

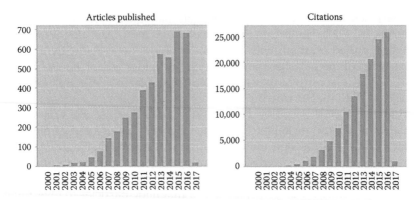

FIGURE 1.3
Number of articles published and number of citations since 2000 for the search term "magnetic + nanoparticle + biomedical." (From Web of Science, Jan. 2017.)

two-dimensional (2D) and 3D cell patterning, including via "magnetic levitation," for applications in tissue engineering and drug discovery, as we see in Chapters 8 and 10. Such 3D cell constructs have many potential applications not only in tissue engineering and regenerative medicine but also in drug screening.

In addition to the work described in this book, there are several very new techniques being developed that rely on the interactions of magnetic fields with magnetic particles and structures inside the body, such as magnetic particle imaging.[24] Indeed, the use of MNPs in biomedicine is expanding at a rapid pace since the early work in this field in the 1990s (Figure 1.3). The ability to remotely manipulate or control actuators and sensors inside the body using magnetic fields outside the body presents myriad possibilities for the development of biomedical technologies aimed at understanding and treating disease. This book explores some of these current technologies.

References

1. Creixell, M., Bohorquez, A.C., Torres-Lugo, M., and Rinaldi, C. (2013). EGFR-targeted magnetic nanoparticle heaters can kill cancer cells without a perceptible temperature rise. *ACS Nano* 5: 7124–7129.
2. Kozissnik, B., Bohorquez, A.C., Dobson, J., and Rinaldi, C. (2013). Magnetic fluid hyperthermia: Advances, challenges and opportunity. *Int. J. Hyperthermia* 29: 706–714.
3. Pankhurst, Q.A., Connoly, J., Jones, S.K., and Dobson, J. (2003). Applications of magnetic nanoparticles in biomedicine. *J. Phys. D: Appl. Phys.* 36: R167–R181.
4. Pankhurst, Q.A., Thanh, N.K.T., Jones, S.K., and Dobson, J. (2009). Progress in applications of magnetic nanoparticles in biomedicine. *J. Phys. D: Appl. Phys.* 42: 224001 (15 pp).

5. Jiles, D. (1998). *Introduction to Magnetism and Magnetic Materials*. Boca Raton, FL: Chapman & Hall/CRC, 536 pp.
6. Craik, D. (1995). *Magnetism: Principles and Applications*. Chichester, UK: John Wiley & Sons, 459 pp.
7. Heilbrunn, L.V. (1922). Eine neue Methode zur Bestimmung der Viskosität lebender Protoplasten. *Jahrb. Wiss. Bot.* 61: 284.
8. Seifritz, W. (1924). An elastic value of the protoplasm with further observations on the viscosity of the protoplasm. *Brit. J. Exper. Biol.* 2: 1–11.
9. Crick, F.H.C., Hughes, A.F.W. (1950). The physical properties of cytoplasm: A study by means of the magnetic particle method. *Exp. Cell Res.* 1: 37–80.
10. Valberg, P.A., Albertini, D.F. (1985). Cytoplasmic motions, rheology and structure probed by a novel magnetic particle method. *J. Cell Biol.* 101: 130–140.
11. Valberg, P.A., Feldman, H.A. (1987). Magnetic particle motions within living cells. Measurement of cytoplasmic viscosity and motile activity. *Biophys. J.* 52: 551–569.
12. Valberg, P.A., Butler, J.P. (1987). Magnetic particle motions within living cells. Physical theory and techniques. *Biophys. J.* 52: 537–550.
13. Wang, N., Butler, J.P., and Ingber, D.E. (1993). Mechanotransduction across the cell surface and through the cytoskeleton. *Science* 260: 1124–1127.
14. Wang, N., Ingber, D.E. (1995). Probing transmembrane mechanical coupling and cytomechanics using magnetic twisting cytometry. *Biochem. Cell Biol.* 73: 327–335.
15. Ingber, D.E. (2000). Mechanical control of cyclic cAMP signalling and gene transcription through integrins. *Nat. Cell Biol.* 2: 666–668.
16. Pommerenke, H., Schreiber, E., Durr, F., Nebe, B., Hahnel, C., Moller, W., and Rychly, J. (1996). Stimulation of integrin receptors using a magnetic drag force device induces intracellular free calcium response. *Eur. J. Cell Biol.* 70: 157–164.
17. Bausch, A.R., Hellerer, U., Essler, M., Aepfelbacher, M., and Sackmann, E. (2001). Rapid stiffening of integrin receptor-actin linkages in endothelial cells stimulated with thrombin: A magnetic bead microrheology study. *Biophys. J.* 80: 2649–2657.
18. Bausch, A.R., Moller, W., and Sackmann, E. (1999). Measurement of local viscoelasticity and forces in living cells by magnetic tweezers. *Biophys. J.* 76: 573–579.
19. Glogauer, M., Ferrier, J., and McCulloch, C.A.G. (1995). Magnetic fields applied to collagen coated beads induce stretch-activated Ca^{2+} flux in fibroblasts. *Am. J. Physiol.* 38: C1093–C1104.
20. Glogauer, M., Ferrier, J. (1998). A new method for application of force to cells via ferric oxide beads. *Eur. J. Physiol.* 435: 320–327.
21. Cartmell, S.H., Dobson, J., Verschueren, S., and Haj EL, A.J. (2002). Development of magnetic particle techniques for long-term culture of bone cells with intermittent mechanical activation. *IEEE Trans. Nanobioscience* 1: 92–97.
22. Dobson, J. (2008). Remote control of cellular behavior with magnetic nanoparticles. *Nature Nanotechnology* 3: 139–143.
23. Bin, H., Haj EL, A.J., Dobson, J. (2013). Receptor-targeted, magneto-mechanical stimulation of osteogenic differentiation of human bone marrow-derived mesenchymal stem cells. *Int. J. Mol. Sci.* 14: 19276–19293.
24. Weizenecker, J., Gleich, B., Rahmer, J., Dahnke, H., and Borgert, J. (2009). Three-dimensional real-time in vivo magnetic particle imaging. *Phys. Med. Biol.* 54: L1–L10.

2

Synthesis and Surface Functionalization of Ferrite Nanoparticles

Jennifer S. Andrew, Carlos Rinaldi, and O. Thompson Mefford

CONTENTS

2.1 Ferrite Crystal Structure 11
2.2 Aqueous Coprecipitation 13
2.3 Thermal Decomposition 13
2.4 Surface Modification.. 16
2.5 Inorganic Modification.. 17
2.6 Organic Modification 18
2.7 Anchoring Groups 19
2.8 Interparticle Interactions 21
2.9 Characterization of the Colloidal State 22
    2.9.1 Dynamic Light Scattering.. 22
    2.9.2 Small Angle X-Ray Scattering 23
    2.9.3 Small Angle Neutron Scattering.. 23
2.10 Additional Modification 24
2.11 Summary and Outlook 26
References.. 27

The past two decades have seen a tremendous amount of progress in the synthesis and application of inorganic nanoparticles. Research in this area has been driven by the unique properties of nanoparticles to their bulk counterparts. A significant portion of this research has been motivated by a desire to control the nanoparticle physical, chemical, and electronic properties through their size and shape. As a result, we can currently synthesize a wide range of inorganic nanoparticles (e.g., metal, metal oxide, semiconductor, etc.) with exquisite control over size (from single nm to 100s of nm) and shape (e.g., spheres, rods, nanoparticles, etc.).[1–28]

Magnetic nanoparticles are one scientifically and practically significant example of the unique behavior and applications made possible at the nanoscale. Depending on their size, shape, and composition, magnetic nanoparticles find use in applications such as magnetic data storage, electromagnetic shielding, magnetically recoverable catalysts, separations, magnetic resonance contrast agents, vehicles for magnetic drug targeting, magnetofection agents, and thermal cancer therapies.[29–36] Ferrites (MFe_2O_4, M = Fe, Co, Ni, Mn) are the most commonly used magnetic nanoparticles and can be obtained through a large number of synthetic strategies. These strategies can be broadly classified into two categories: (1) metal salt reduction and coprecipitation methods that generally yield particles with poor control over size and shape[37–42] and (2) thermal decomposition reactions using organometallic precursors that can yield particles with exquisite control over size and shape.[2,13,26,43,44]

The synthesis of magnetic nanoparticles has been an active area of research since the development of *ferrofluids* in the 1950s and 1960s.[33] As with many other inorganic nanoparticles, methods for the synthesis of magnetic nanoparticles are generally categorized as either "top-down" or "bottom-up" approaches. Top-down methods involve starting with bulk material and achieving nanoscale features through mechanical or chemical processes. A traditional top-down approach involves high-energy ball-milling of bulk material until particle dimensions reach the nanoscale. Although this process is well suited to producing large quantities of material, it is plagued by quality control issues such as impurities from the milling media and poor control over the shape and size dispersion of the resultant nanoparticles.[45–47] Lithographic techniques represent an alternative top-down approach that affords excellent control over shape and size dispersion, but is limited to particles with diameters greater than 50 nm and have limited scalability.[48–53]

As a result of the limitations of top-down approaches, the vast majority of research on the synthesis of nanoparticles has focused on bottom-up methods, which can be further classified as gas-phase[54–57] or wet-chemistry approaches.[14,18,19,33,37,41] Gas-phase synthesis methods include cluster-deposition, which allows for excellent control of nanoparticle size and composition but has low yield and flame pyrolysis, which can produce large quantities of particles in continuous operation, but lacks fine control over nanoparticle shape and size. Furthermore, gas-phase synthesis methods face significant problems in subsequent postsynthesis modification because the absence of a nanoparticle-capping agent during synthesis leads to their aggregation. Once formed, these aggregates can be quite difficult to break and ultimately limit the properties and applications of the nanoparticles.

Wet-chemistry approaches to nanoparticle synthesis afford great flexibility in the choice of precursors used to generate the particles (most commonly metal salts or organometallic complexes), reaction medium (aqueous or organic solvents, or supercritical fluids), nature of capping agents (from simple surfactants to polymers or even biomolecules). Additionally, additives can be used to control particle shape, and these methods include other variables that can be tuned to vary the properties of the resultant particles, such as reaction temperature, reagent addition rates, and concentrations. The presence of a capping agent in wet-chemistry approaches makes these particles amenable to postsynthesis surface modification, providing additional tunability of the final properties or the nanoparticles, making them useful in many applications.

In this chapter, we focus on the synthesis, characterization, and surface functionalization of ferrite nanoparticles. The following topics are presented:

- An overview of the spinel crystal structure
- Methods and theory to synthesize magnetic ferrite nanoparticles
- Methods to surface functionalize ferrite nanoparticles

2.1 Ferrite Crystal Structure

Prior to discussing the synthetic methods that have been used to produce ferrite nanoparticles, it is important to understand the crystal structure of ferrites and how this structure determines their ultimate magnetic properties. Most ferrite materials form in the spinel structure, AB_2O_4. In the case of ferrites, the B cation is Fe^{3+} and the A cation is a divalent transition metal such as Fe^{2+}, Co^{2+}, or Ni^{2+}. The spinel structure, shown in Figure 2.1, is defined by a face-centered cubic lattice of oxygen anions, where the metal cations are distributed on the tetrahedrally or octahedrally coordinated sites, referred to as the A and B sites, respectively. Each unit cell contains eight AB_2O_4 formula units, with 32 oxygen anions, 16 trivalent cations, and 8 divalent cations. The distribution of the cations on the A and B sites determines whether the spinel is normal or inverse. In a normal spinel all the trivalent cations sit on the octahedral B sites and all the divalent cations sit

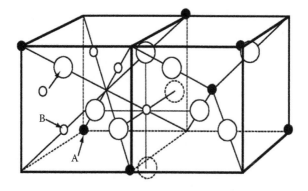

FIGURE 2.1
Two octants of the spinel unit cell. The large spheres represent the oxygen anions, while the small black and white spheres represent the metal cations in the tetrahedral and octahedral sites, respectively. The spinel unit cell is made up of eight of these alternating octants. (From LaMer, V. K.; Dinegar, R. H., *Journal of the American Chemical Society*, 72, 4847–4854, 1950.)

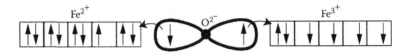

FIGURE 2.2

Schematic of the superexchange mechanism between octahedrally coordinated Fe^{2+} and tetrahedrally coordinated Fe^3 in Fe_3O_4.

on tetrahedral A sites. For the inverse spinel, the divalent cations share the octahedral B sites with one half of the trivalent cations. The other half of the trivalent cations sit on the tetrahedral A sites. The normal and inverse spinel structures represent the two extremes of cation distribution in spinels, and in reality, the cations can be distributed in any manner within these two forms, depending on the ionic sizes of the cations and processing conditions.

In ferrites, the magnetic properties are determined by the distribution of the cations in the octahedral and tetrahedral sites. The magnetic cations order antiferromagneticaly via superexchange, which is dominated by the exchange interaction between the octahedrally and tetrahedrally coordinated cations. Figure 2.2 shows a schematic of this interaction, for the case of Fe_3O_4, between Fe^{2+} and Fe^{3+}.

Fe_3O_4 has the inverse spinel structure, where all the Fe^{2+} cations sit on octahedral sites, while half of the Fe^{3+} cations on octahedral sites and half on tetrahedral sites. The overall magnetization can be calculated as the difference in magnetization of the octahedral and tetrahedral lattices.[58] In nickel ferrite, the magnetic moment of the tetrahedral ions is 5 μ_B (Bohr magnetons), the magnetic moment of Fe^{3+}, and the magnetic moment of the octahedral ions is 7 μ_B, the sum of the magnetic moments of Ni^{2+} and Fe^{3+}. Therefore, the overall magnetic moment per molecule of $NiFe_2O_4$ is 2 μ_B. To convert between Bohr magnetons and units that are more traditionally used in the literature, one can use the conversion that $\mu_B = 9.274 \times 10^{-24}$ J/T, where 1 J/T is also equivalent to 1 Am^2. As outlined above, each spinel unit cell contains eight formula units of the form MFe_2O_4, so for the case of $NiFe_2O_4$, one $NiFe_2O_4$ unit cell would contain eight formula units of $NiFe_2O_4$ and therefore 16 μ_B. By taking the above conversions alongside the lattice parameters and density of bulk ferrite materials, the magnetization in more conventional A/m or emu/g can be obtained. It is of course worth noting that the lattice parameters, and thereby the density of oxide nanoparticles do tend to deviate from bulk values adding additional uncertainty to these calculations. Lastly, these are the theoretically determined values for the Bohr magnetons per unit cell, and often in reality, these can vary in real-world materials depending on the distribution of cations between the octahedral and tetrahedral sites.

2.2 Aqueous Coprecipitation

Aqueous coprecipitation is one of the more commonly used wet-chemistry methods for the synthesis of ferrite nanoparticles of the form MFe_2O_4, where M = Mn, Fe, Co, Ni, Zn. The following discussion focuses on the synthesis of Fe_3O_4; however, similar results can be obtained for other metal ferrites, MFe_2O_4. The aqueous coprecipitation technique is based on a precipitation model developed by LaMer for the condensation of monodispersed hydrosols from a supersaturated solution.[59]

In the case of aqueous synthesis, magnetic nanoparticles are typically obtained by the coprecipitation of metal salts in an alkaline environment.[38–42] This reaction follows the overall stoichiometry:

$$M^{2+} + 2Fe^{3+} + 8OH^- \rightarrow MFe_2O_4 + 4H_2O \qquad (2.1)$$

where M = Co, Ni, Fe, Mn. It is worth noting, however, that the initial precipitate in this reaction is not the ferrite, but are instead metal hydroxides that with time at an elevated temperature form the resultant oxide via a condensation process, in which the cations are linked by oxygen ions.[37] Thus, Equation 2.1 is really an overall representation for a complex cascade of chemical steps that ultimately yield the desired ferrite. Aqueous coprecipitation methods have several advantages, including the lack of toxic solvents, comparatively low temperatures (~80°C–90°C usually), moderate yields, and the generation of water-soluble particles through the appropriate use of capping agents. However, particles produced by this method lack uniformity in size and shape, have high fractions of small aggregates (even with the use of capping agents), and tend to have magnetic properties that are significantly lower than bulk values.

2.3 Thermal Decomposition

To overcome the limitations of aqueous coprecipitation, much work has focused on the synthesis of ferrite nanoparticles via thermal decomposition of organometallic precursors in the presence of a surfactant capping agent.[13,18–20,33] These methods yield particles with narrow size distributions and with excellent control over size and shape at laboratory scale syntheses of ~4 g or less. The LaMer model for precipitation of an insoluble colloid from a supersaturated solution,[59] illustrated in Figure 2.3, can also be used to explain the success of thermal decomposition routes to nanoparticle synthesis. According to LaMer, initially, the concentration of

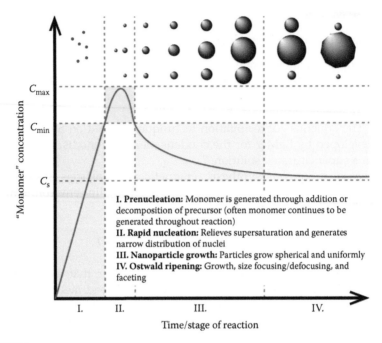

FIGURE 2.3
LaMer model of precipitation and growth of an insoluble phase from an initially supersaturated solution.

the nanoparticle precursor increases until it reaches some critical value at which nucleation of the insoluble product occurs spontaneously and continues as long as it is above the saturation concentration. After this rapid nucleation step, there is a regime of diffusion-controlled growth that occurs until the solubility limit is reached. Once the solubility limit is reached, there is no longer a driving force for precipitation of the insoluble phase. The key to forming nanoparticles with narrow size distribution is thus to generate a single, short nucleation event followed by controlled particle growth. This is difficult to achieve in aqueous phase precipitation but feasible with thermal decomposition through selection of organometallic precursors with thermal degradation profiles compatible with the reaction temperature.[13,15,28,60–62]

Thermal decomposition syntheses can in turn be classified based on the organometallic precursor used and whether or not they are "heated-up" with the solvent or "hot-injected" when the solvent has reached an appropriate temperature. Iron acetyl acetonates [Fe(acac)$_3$] and iron pentacarbonyls [Fe(CO)$_5$] are typically used in "hot-injection" reactions,[19,20,63,64] where the objective is to rapidly inject the precursor into a high-temperature solvent to induce rapid nucleation. On the other hand, in the "heating-up" method developed by Hyeon and collaborators,[13,14,65,66] an iron oleate [Fe(OA)$_3$, where OA = CH$_3$(CH$_2$)$_7$CH=CH(CH$_2$)$_7$COO⁻], is brought to the reaction temperature at a constant rate. As the temperature rises, the iron oleate undergoes a series of decomposition steps, thus the approach to supersaturation of the nanoparticle precursor and the nucleation event is controlled through the temperature ramp rate.[65] In both of these methods, "hot-injection" and "heating-up," once nuclei have formed nanoparticles, growth occurs due to incorporation of iron that remains in solution after the initial nucleation event, until the overall concentration reaches the saturation value. After this, further size-focusing can occur due to Ostwald ripening, which unfortunately can also lead to faceting of the nanoparticles and in some cases size-defocusing.[67] Both reactions are typically carried out under reflux, with final nanoparticle size controlled through solvent selection (i.e., reaction temperature)[13,19] and by changing the concentration of precursors and using different surfactants or surfactant mixtures (e.g., oleylamine, oleic acid, trioctylphosphine oxide [TOPO], etc.).[68] Of these two general thermal decomposition approaches, the "heating-up" method using iron oleates has recently become the most popular as it allows for synthesis at a relatively larger scale of ~4 g of the inorganic core and avoids problems with temperature control and off-gas formation inherent to the "hot injection" method. Further, a variety of ferrite compositions have been obtained by these methods through the use of combinations of organometallic precursor.[18,66]

The development of thermal decomposition methods has enabled fine control over nanoparticle crystal size, shape, and composition. This has in turn enabled studies of the effect of size on physical and magnetic properties of nanoparticles and the development and testing of magnetic nanoparticles for a variety of applications.[32,34,36,37,69,70] Still, thermal decomposition methods suffer from their own limitations. For example, currently, changing the size of the nanoparticles is accomplished by changing the nature of solvent (i.e., reaction temperature) and surfactants (Figure 2.4). The specific discovery of conditions that yield a desired size typically occurs through heuristics or trial-and-error. Further, studies comparing particles of different sizes typically obtain these under widely different conditions (reaction temperature, nature of surfactant, identity of organometallic precursor, etc.) and it is unclear how such conditions can affect and therefore confound the study. This is perhaps one reason why detailed studies of the effect of size on magnetic properties of nanoparticles obtained by thermal decomposition routes remain limited. Furthermore, it is becoming increasingly apparent that these methods suffer from significant batch-to-batch and lab-to-lab variability in particle size and properties. That is, one can run the reaction under "identical" conditions multiple times and obtain batches that each have narrow size distributions but vary significantly (relative to their size distribution) in terms of mean size.

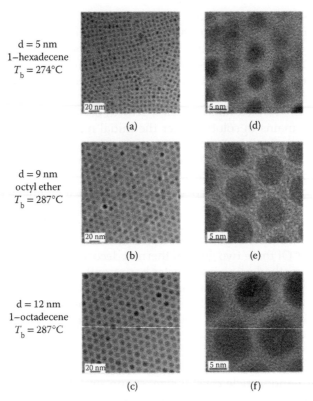

FIGURE 2.4

Transmission electron microscopy (TEM) and high-resolution transmission electron microscopy (HRTEM) images as a function of solvent. (Reprinted with permission from Nature Publishing Group.)

2.4 Surface Modification

Magnetic particles for biomedical applications have to fulfill a whole range of criteria.[71] In most cases, these particles should be stable under biological conditions, they should be biocompatible, and they should have a hydrodynamic diameter between 20–60 nm in order to avoid renal clearance and/or elimination through the reticuloendothelial system (RES) ensuring a longer blood half-life. Additionally, functional groups on the surface of the magnetic particles are important to enable the attachment of biological ligands. The coating material chosen for these particles determines most of these characteristics. Here, we discuss the most prevalent coatings for magnetic nanoparticles in biomedical applications.

Typically, following the synthesis of magnetic nanoparticles, additional surface modification is necessary to provide the following:

1. Long-term colloidal stability via controlled interparticle interactions for sufficient shelf life and surface chemistry, which will stay bound to the particle in biological media

2. A biocompatible surface that will minimize the risk of an immune response

3. Additional functionality for imaging, therapy, and targeting

Despite some efforts to produce magnetic nanomaterials that meet these criteria in a single step or "one pot" reaction,[72-74] the majority of these materials are formed by completing a particle synthesis step with a sacrificial ligand followed by surface modification. For example, as discussed above, magnetic particles with narrow size distributions can be produced via the thermal decomposition of chelated metal precursors. The resulting particles are hydrophobic, typically stabilized with a fatty acid, such as oleic acid. Prior to use in biological systems, the hydrophobic surface of these particles needs to be altered for use in aqueous systems.

This modification can be separated into two main categories: inorganic and organic. While inorganic coatings provide particles with a shell, organic coatings, depending on the polymer used, can provide a matrix surrounding the particle core. Even though there is growing interest in core shell particles using different ceramic coatings, for example, $CoFe_2O_4$-coated $MnFe_2O_4$ for magnetic fluid hyperthermia,[75] the two most popular inorganic coatings are still silica and gold.

2.5 Inorganic Modification

Silica stabilizes the particles by reducing magnetic interactions between particles, but also provides a negative surface charge.[76] Amorphous silica (SiO_2) can be deposited through the base-catalyzed hydrolysis of tetraalkyl orthosilicate, such as tetraethyl orthosilicate (TEOS) in an ethanol/water mix.[77] However, its growth is often hard to control. Therefore, often a templating agent such as cetyltrimethylammonium bromide (CTAB) is introduced before the TEOS, generating a mesoporous silica shell containing 2–4 nm pores, allowing the loading of hydrophobic drugs such as cyclosporin A.[78] Other silicon-ligands enabling an amorphous silica shell are trialkoxysilylpropanes, such as (3-aminopropyl)triethoxysilane (APTES) or (3-mercaptopropyl)triethoxysilane (MPTES), bearing a functional group.[79]

Gold is another common inorganic coating, as it enables the attachment of thiol-containing ligands, such as 2-mercapto-ethylamine, cystamine, thio-acetic acid, or dithiobis(succinimidyl)propionate (DSP).[80] Particles can be coated by gold using sacrificial reduction, where the outer layer of the core particle is oxidized, acting as a reducing agent for the Au^{3+} ions.[79]

2.6 Organic Modification

Organic surface modification includes ligand exchange with a more favor-able ligand that can meet the specific needs of applications, with polymers being the most common. Other reagents include 2,3-dimercaptosuccinnic acid (DMSA),[81–83] or oxidation of the oleic acid to produce negatively charged particles.[11,84] Examples of polymers that have been used to modify the sur-face include dextran and cyclodextran,[85–94] polydimethylsiloxane (PDMS),[95–98] poly(N-isopropylacrylamide) (PNPAM),[99–103] chitosan,[104–108] poly(glycidyl methacrylate) (PGMA),[109] and polyethylene oxide (PEO), also known as poly-ethylene glycol (PEG).[90,110–119]

Dextran, a polysaccharide composed of multiple units of α-D-glucopyranosyl, used as a plasma expander in the clinic is most often used for its biocompat-ibility and low cytotoxicity. Dextran is either adsorbed or anchored (through chemical modification) onto the surface of the particle cores. Therefore, the stability of these particles is highly dependent on the affinity of the anchor group on the dextran chain to the particle core, as well as molecular weight of the dextran used. Dextran derivatives, like carboxydextran, have been used to coat clinically approved magnetic resonance imaging (MRI) contrast agents, such as Resovist®[120] and Endorem®.[121]

Chitosan is a linear biocompatible and biodegradable polysaccharide, syn-thesized from chitin. In addition to hydroxyl groups, the structure of chito-san provides amine groups for anchoring or functionalization. These amine groups are also responsible for an overall positive charge, enabling not only the interaction with negative hydroxyl groups (Fe-OH) on the surface of magnetite cores, but also association with DNA for gene transfection.[122–124]

PEO or PEG is one of the most popular choices. PEO is also soluble in water and many organic solvents, which allows for a variety routes in chemi-cal modification of its end groups.[125] Because of its solubility in water and capability of modification, PEO is ideal for biomedical applications; it has low immunogenity, renal clearance in vivo, and is not biodegradable.[126] Due to its zero net charge, it prevents the adsorption of proteins and therefore extends the blood half-life of magnetic nanoparticles, enabling the genera-tion of truly "stealth" nanoparticles. Furthermore, PEG is amphilic, soluble in water, as well as a range of organic solvents, such as ethanol, acetone or chloroform, making it a popular spacer molecule.[127,128] The polymer has been FDA approved for use in a variety of cosmetics, pharmaceuticals, and even an additive food.[129] PEO is ideal for theranostic systems, not only because of how inert the backbone is to protein absorption,[130] but this relatively unreactive polymer chain allows for modification of the end groups with relative ease.[125,131]

2.7 Anchoring Groups

In addition to having a biocompatible stabilizing layer, it is also important to have a good anchor on the surface of the magnetic particle. This anchor needs to be able to strongly bind to the surface of the particle and remain there throughout the lifetime of its use. Biological media is especially challenging due to the high ionic strength, physiological pH, protein fouling, and elevated temperatures.[132] Coatings lacking strong anchoring groups, such as unmodified dextran and chitosan, are typically high-molecular-weight (>10 kDa) polymers, and predominantly adsorb onto the surface of the nanoparticle core without a specific anchor group. This leads to particles that are stable in water, but not under biological conditions.[93]

Alternatively, lower molecular weight ligands in combination with robust anchors that couple irreversibly to the nanoparticle core have the potential to result in more stable particles in physiological media. Many different anchoring groups have been investigated.[80,133] Most notable are carboxylates, amines, thiols, silanes, phosphonates, and catechols (Figure 2.5).

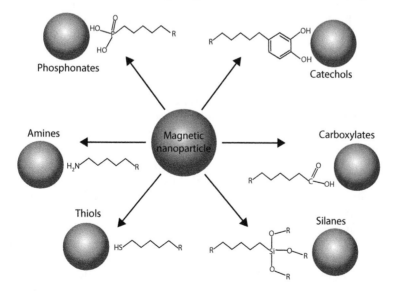

FIGURE 2.5
Cartoon representing different anchoring chemistries used to bind to the surface of magnetic nanoparticles.

With each of these systems, the central goal is to strongly bind to the surface of the magnetic particle in a manner that is stable in media (i.e., will not come off of the surface during storage or during the biomedical application).

Another benefit of the low-molecular-weight stabilizers is that better coverage of the core can be achieved through higher packing densities, further improving stability.[134] In 2009, Goff et al.[135] compared PEO oligomers carrying different functional groups (a single carboxylate, a single ammonium, a single phosphonate zwitterion, and three phosphonate zwitterions). The results showed that while carboxylate and ammonium groups allowed the polymer to desorb from the magnetite core, the phosphonate PEOs produced stable particles. A recent comparison of amines, carboxylates, phosphates, and two catechol-based PEO ligands demonstrated that amines and carboxylates are poor at displacing oleic acid from the surface of particles produced via thermal decomposition. However, phosphates and catechols demonstrated the highest rate of exchange.[136] Amstad et al.[137] performed a comparison of catechol derivative-anchored ligands. They compared three anchors: catechol, mimosine, and nitrocatechol. While the catechol stayed only weakly interacted with the magnetite core, the mimosine's high affinity actually removed Fe^{3+} ions from the core slowly dissolving it. However, the addition of an electronegative group, such as NO_2, to the catechol ring, as in the case of the nitrocatechol, was shown to enhance the interaction between anchor and iron oxide and therefore the particle stability.

An issue with monofunctional anchor groups is competition from naturally occurring ligands. For example, carboxylate and even monofunctional catechol-based ligands have been demonstrated to be removed from the surface of particles in phosphate buffered saline (PBS).[136,138] It is hypothesized that the phosphates in the solution exchange with the ligands on the surface of the particles. To counter this, "multidentate" or "multivalent" approaches have been proposed. Here, each stabilizing ligand contains multiple anchoring groups. It has been proposed that the presence of two or more anchors results in faster exchange kinetics, and greater long-term stability.[139] For example, Na et al. used a polyacrylic acid-based backbone to which they bonded catechol anchoring groups and PEO brushes. The resulting particles had strong colloidal stability in a wide range of pH and no toxicological effects.[140] In addition to catechol-based multi-anchored systems, phosphonate systems have also been demonstrated.[136,138–141] Alternatively, polymers such as polyethylenimine (PEI) can be used to functionalize the surface where the repeat unit contains a specific binding group, and thus provides multiple anchoring points.[142] This has been shown to provide good stability, and can be additionally functionalized with polymers such as PEG for greater stability.[143] However, the resulting highly charged surface might be of concern with biofouling.[88]

2.8 Interparticle Interactions

Central to the success of these materials in biomedical applications is the control of interparticle interactions in biological media. These interactions control the overall colloidal stability of these systems, and thus have implications on the long-term shelf life of these materials as well as effecting their biodistribution, immune response, and potential for targeting and therapy. One approach to calculate these interactions is the summation of the van der Waals, electrostatic, steric, and magnetic interactions, through what is commonly known as the Derjaguin–Landau–Verwey–Overbeek (DLVO) theory of colloidal stability.[144]

Van der Waals forces represent attractive and repulsive forces that are greatest in magnitude in short-range distances (typically <2 nm). Van der Waals forces include several different types of interactions, including the Keesom force between permanent dipoles, the Debye force between induced dipoles, and London dispersion forces.

Electrostatic interactions represent interactions between charged surfaces and are heavily influenced by the surface charge of the particles and the properties of the media. For example, negatively charged particles can be easily dispersed in deionized water, but the introduction of ions to the media can result in "screening" of the surface charge, and this can in turn result in aggregation of the particles. For biomedical applications, where the ionic concentration of media is relatively high, relying solely on electrostatic interactions can be challenging. Zeta potential measurements are commonly used to determine the magnitude of electrostatic potential energy.

Steric interactions are repulsive forces that result from space filling, excluded volume, and osmotic pressure buildup when two particles approach each other. These can therefore by used to prevent aggregation. As discussed above, surface modification with a polymer brush can provide the repulsive forces necessary for long-term colloidal stability. Dynamic light scattering (DLS) and other scattering methods, as discussed in Section 2.9, can be used to determine the hydrodynamic diameter of these systems and evaluate changes in colloidal stability over time.

Finally, magnetic interactions of the particles need to be accounted for. An initial first approximation to calculate these interactions is treating each particle as a point dipole. However, this must be done with a few caveats. First, the size of the magnetic particle must be considered. As discussed in Chapter 1 monodomain particles below a key size will randomly orient their magnetic spins due to thermal fluctuations. Magnetometry will indicate zero net moment in the absence of an applied field. However, each particle retains a rapidly fluctuating magnetic moment. When two particles approach each other, these fluctuating magnetic moments can couple, leading to magnetic interactions between the particles. Thus, dipole interactions of these materials must still be considered. A typical conservative approach is to consider the case of highest interactions. This is when the moments of two particles are aligned, resulting in the greatest magnitude of attractive interactions. In applied fields, this alignment is greater, thus the likelihood of increased interparticle interactions will increase, thus agglomerations of these materials is more likely.

2.9 Characterization of the Colloidal State

Most techniques used to characterize the colloidal state of magnetic particles utilize some sort of scattering technique, where a sample is probed by measuring the scattering from a probing electromagnetic wave. The wavelength of the incident electromagnetic wave will dictate the resulting scattering length, and thus the scale at which information can be obtained. In addition, different wavelengths (λ) interact with materials in different ways. For example, photons ($\lambda \approx 500$ nm) interact with the electron structure of atoms, x-rays ($\lambda \approx 0.15$ nm) are scattered by the electron cloud, and neutrons ($\lambda \approx 0.4$ nm) by the nuclei of atoms.[145] Below are brief discussions of scattering with each of these sources. The reader is encouraged to seek more detailed texts for greater discussion.

2.9.1 Dynamic Light Scattering

Also known as photon correlation spectroscopy (PCS) or quasi-elastic light scattering (QELS), DLS is one of the most common techniques to characterize magnetic nanoparticles. Specifically, this technique measures the hydrodynamic size of the particles, which includes the particle core, any surface modification, and the associated ion cloud surrounding the complex. The technique actually measures the Brownian motion of the particles in suspension and uses an algorithm to determine size based on the estimated translational diffusivity of the particles and the Stokes–Einstein relation.[146] This technique is especially useful for the detection of aggregation or agglomeration of the particles, as DLS is much more sensitive to larger objects. Most instruments use a single monochromatic laser that passes through the colloidal suspension of particles. The scattering of light from the suspension is then detected by a single detector at a fixed angle, a detector on a goniometer for multiple angle detection, or multiple detectors at fixed angles.[147] This scattered light is the result of constructive and destructive interference based on the motion of particles relative to the laser's path. This interference results in fluctuations in the scattering as a function of time, which is measured by the detectors. Correlation functions are then used to interpret the changes in intensity of scattering as a function of time. These correlation functions are then fitted with an algorithm to typically produce log-normal based distributions of the hydrodynamic size of the particles. A user interpreting data should be concerned about the weighting of these distributions. For example, intensity-based scattering is based on the radius of the particle to the sixth power, whereas number average is to the first (volume is r^3). Thus, intensity-weighted distributions are more sensitive to the presence of aggregates than number-weighted distributions. Importantly, the user must keep in mind that the intensity-weighted distribution is directly determined from the measurements, whereas other distributions are calculated and are therefore more susceptible to error.

2.9.2 Small Angle X-Ray Scattering

Small angle x-ray scattering (SAXS) is an effective technique to determine the size of the inorganic core of a particle. As heavier elements are more likely to scatter x-rays, SAXS measurements are heavily influenced by heavier elements and not organics. Despite this, much information regarding the size of the particles,[148] as well as the structure of the materials in the suspension[149] including chaining[150] and clusters of particles.[151] Composites consisting of magnetic particles in an organic matrix have also been characterized by SAXS to measure the distribution of the particles within the matrix.[152,153] This technique is also effective in measuring assemblies of particles.[154-156] For example, clusters of magnetic nanoparticles for use at T2 MRI contrast agents were characterized via SAXS to determine cluster size[151] and then calculate theoretical calculations based on SAXS-T2 relationships established by Carroll et al.[157]

2.9.3 Small Angle Neutron Scattering

Small angle neutron scattering (SANS) is a powerful technique for probing the organic stabilizing layer on the inorganic particle. Often, this is accomplished by contrast matching with the appropriate deuterated solvent to allow for better scattering of the material of interest. Several researchers have utilized this technique to determine the thickness of the stabilized polymer brush as a function of solvent or thermal environment.[158-160] More recently, the spin structure of the magnetic particles have been investigated.[154] This has included the use of a polarized source of neutrons that demonstrated spin canting on the surface of 9-nm magnetite nanoparticles in a 1.2 T field, where the thickness of this canted shell was dependent on temperature and applied field.[161] Nonetheless, the major drawbacks to SANS is the need for experiments to be conducted near a steady source of neutrons. Typically, these are research reactors. In addition, the number of counts needed for each experiment can limit the observation of kinetic activities.

2.10 Additional Modification

Often a research team is challenged to provide properties in addition to those of the magnetic core. These additional properties can include added biofunctionality (including targeting and therapeutic agents), optical tags, and polymeric-based stabilizing compounds.[162,163] There are multiple tools available to link additional functionality. The reader is encouraged to read many other available texts that address these reactions in greater detail.[164,165]

One of the most widely used conjugation techniques is the linking of amines with carboxylates to form amide bonds, that is, amidation. The ability to react with an amino group is an important tool for modification of magnetic nanoparticles. Amines can be found in surface modification polymers such as PEI and amino-terminated PEG. Also, amines can be found in a variety of biomolecules including peptides, DNA, antibodies, etc.[164] Amidation can be facilitated by a number ways (Figure 2.6) including coupling of carboxylic acids with amine (typically using 1-ethyl-3-[3-dimethylaminopropyl] carbodiimide [EDC]). More often, EDC is used

Amidation schemes

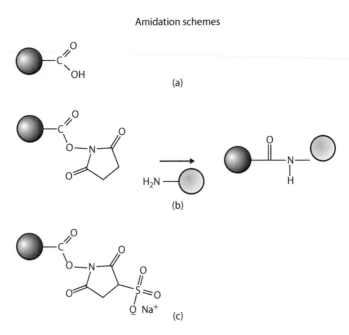

FIGURE 2.6
Illustration of various routes to form amide bonds including: (a) carboxylic acid, (b) NHS activated, and (c) sulfo-NHS activated.

in combination with either *N*-hydroxysuccinimide (NHS) for organic reactions or *N*-hydroxysulfosuccinimide (sulfo-NHS) for aqueous synthesis. Example reactions with magnetic particles include folic acid[166] and aptamers.[167]

"Click" reactions represent one of the most popular versions of bioorthogonal reactions. The reaction that is almost synonymous with "click chemistry" is the 1,3-dipolar cycloaddition of alkynes and azides yielding an triazole developed by Huisgen.[168] Alkyne and azide functional groups are extremely stable toward water and oxygen and many synthetic conditions.[169] The mild reaction conditions of the copper-catalyzed reaction of an alkyne and azide make it an ideal synthetic route for bioconjugation. The triazole linkers are ideal for bioconjugation because they are soluble and stable in water, including the reducing nature of biological environments.[170] (See example reaction in Figure 2.6a.) DNA bioconjugation can be quite difficult and inefficient because they require aqueous conditions, Seo et al. were able to tag a strand of DNA, for sequencing studies, modified with an alkyne and with a fluorophore modified with an azide, with a product yield of 91%.[171] A recent comparison of "click" chemistry to carbodiimide-derived functionalization on nanoparticle surfaces demonstrated that "click" chemistry was much more efficient.[172] Unfortunately, the copper catalyst commonly used in this reaction is a toxicity concern for some if it is not properly removed. Fortunately, copper-free reactions have been developed based on the work of Bertozzi et al.[173] These methods are based on straining the alkyne bond, and thus lower the activation energy required to drive coupling with the azide group. (See example reaction in Figure 2.7b.) As an alternative for compounds containing azides, strain-promoted alkyne–nitrone cycloaddition chemistry has been developed (see Figure 2.7c). As an added benefit, these nitrone-based materials can be easily created from oxidation of terminal amines found in many biomolecules.[174]

A large number of reactions are also possible with thiol-terminated groups. Thiols are commonly found in peptides and other biomolecules.[164] Either through free-radical addition or with a catalyst, the conjugation utilizes the relatively weak sulfur–hydrogen interaction and reacts with carbon–carbon double bonds (Figure 2.8a).[175] As a catalyst-free variation, this reaction utilizes the electron-poor double bond of maleimides (Figure 2.8b). This reaction has many advantages apart from being catalyst free; they can be conducted in a variety of solvents and present low toxicity. Amines can also be converted to maleimides by reaction with 6-maleimidohexonaic acid *N*-hydoxysuccinimide ester.[164] Pyridyl disulfide (Figure 2.8c) is also a good thiol-reactive conjugation strategy where pyridine thione is released, resulting in a nearly irreversible reaction.[176]

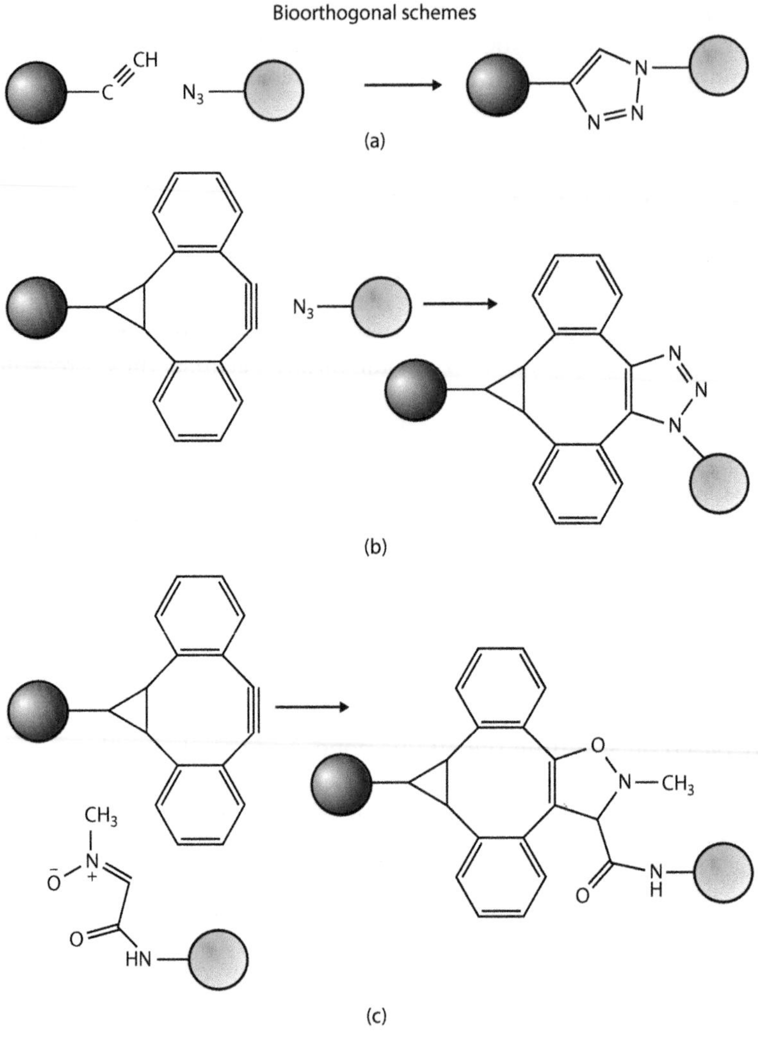

Bioorthogonal schemes

(a)

(b)

(c)

FIGURE 2.7
Representative "click" reactions where (a) is the traditional copper-catalyzed coupling, (b) strain-promoted cycloaddition, and (c) nitrone-alkyne addition.

2.11 Summary and Outlook

This chapter presents an overview of the most widely used synthesis and surface modification methods for ferrite nanoparticles. The intention is to provide the reader with a point of entry into the field. However, it is noted that the literature on the subject is vast, and excellent reviews with recent developments are published on a regular basis. Ultimately, the synthesis and functionalization

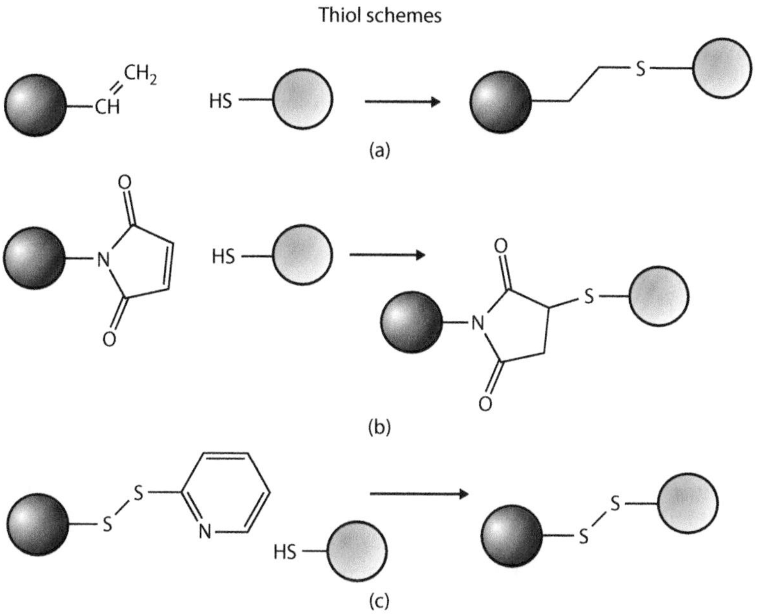

FIGURE 2.8
Schematic of different thiol-based conjugation chemistries where (a) is thiol-ene addition, (b) maleimide reaction, and (c) thiol substitution chemistry.

protocols that are utilized will be influenced by the end application of the magnetic nanoparticles. Though overall crystal quality is lower in the case of aqueous coprecipitation, this route produces water-soluble nanoparticles that can be readily incorporated into biological systems. Thermal decomposition methods provide higher quality and more readily controlled size and shape of the resultant particles, but require postsynthesis modification to impart water solubility. As the synthesis methods and applications of magnetic nanoparticles continue to grow, it is important that these materials be fully characterized to ensure that appropriate structure–property relationships can be realized. These structure–property relationships will ultimately help to determine the reasons behind batch-to-batch and lab-to-lab variability in the properties of magnetic nanoparticles. Through a complete understanding of these structure–property relationships, the application of magnetic nanoparticles in a wide range of medical and electronic applications will be made possible.

References

1. Cheng, G.; Hightwalker, A., Synthesis and characterization of cobalt/gold bimetallic nanoparticles. *Journal of Magnetism and Magnetic Materials* **2007**, *311* (1), 31–35.

2. Cheon, J., Kang, N., Lee, S., Lee, J., Yoon, J., Oh, S., Shape evolution of single-crystalline iron oxide nanocrystals. *Journal of the American Chemical Society* **2004**, *126* (7), 1950–1951.

3. DeVries, G. A., Brunnbauer, M., Hu, Y., Jackson, A. M., Long, B., Neltner, B. T., Uzun, O., Wunsch, B. H., Stellacci, F., Divalent metal nanoparticles. *Science* **2007**, *315* (5810), 358–361.

4. Eustis, S., El-Sayed, M., Why gold nanoparticles are more precious than pretty gold: Noble metal surface plasmon resonance and its enhancement of the radiative and nonradiative properties of nanocrystals of different shapes. *Chemical Society Reviews* **2006**, *35*, 209–217.

5. Gerion, D., Pinaud, F., Williams, S. C., Parak, W. J., Zanchet, D., Weiss, S., Alivisatos, A. P., Synthesis and properties of biocompatible water-soluble silica-coated CdSe/ZnS semiconductor quantum dots. *Journal of Physical Chemistrty B* **2001**, *105* (37), 8861–8871.

6. Hou, Y., Xu, Z., Sun, S., Controlled synthesis and chemical conversions of FeO nanoparticles. *Angewandte Chemie International Edition* **2007**, *119* (33), 6445–6448.

7. Joo, J., Yu, T., Kim, Y. W., Park, H. M., Wu, F., Zhang, J. Z., Hyeon, T., Multigram scale synthesis and characterization of monodisperse tetragonal zirconia nanocrystals. *Journal of the American Chemical Society* **2003**, *125* (21), 6553–6557.

8. Jun, Y. W., Choi, J. S., Cheon, J., Shape control of semiconductor and metal oxide nanocrystals through nonhydrolytic colloidal routes. *Angewandte Chemie International Edition* **2006**, *45* (21), 3414–3439.

9. Kim, D., Lee, N., Park, M., Kim, B. H., An, K., Hyeon, T., Synthesis of uniform ferrimagnetic magnetite nanocubes. *Journal of the American Chemical Society* **2009**, *131*, 454–455.

10. Laurent, S., Forge, D., Port, M., Roch, A., Robic, C., Vander Elst, L., Muller, R. N., Magnetic iron oxide nanoparticles: Synthesis, stabilization, vectorization, physicochemical characterizations, and biological applications. *Chemical Reviews* **2008**, *108* (6), 2064–2110.

11. Lee, S.-Y., Harris, M. T., Surface modification of magnetic nanoparticles capped by oleic acids: Characterization and colloidal stability in polar solvents. *Journal of Colloid and Interface Science* **2006**, *293* (2), 401–408.

12. O'Brien, S., Brus, L., Murray, C. B., Synthesis of monodisperse nanoparticles of barium titanate: Toward a generalized strategy of oxide nanoparticle synthesis. *Journal of the American Chemical Society* **2001**, *123* (48), 12085–12086.

13. Park, J., An, K., Hwang, Y., Park, J.-G., Noh, H.-J., Kim, J.-Y., Park, J.-H., Hwang, N.-M., Hyeon, T., Ultra-large scale syntheses of monodisperse nanocrystals via a simple and inexpensive route. *Nature Materials* **2004**, *3* (12), 891–895.

14. Park, J., Joo, J., Kwon, S. G., Jang, Y., Hyeon, T., Synthesis of monodisperse spherical nanocrystals. *Angewandte Chemie International Edition* **2007**, *46*, 4630–4660.

15. Park, J., Lee, E., Hwang, N. M., Kang, M. S., Kim, S. C., Hwang, Y., Park, J. G., Noh, H. J., Kini, J. Y., Park, J. H., Hyeon, T., One-nanometer-scale size-controlled synthesis of monodisperse magnetic iron oxide nanoparticles. *Angewandte Chemie International Edition* **2005**, *44* (19), 2872–2877.

16. Penner, R. M., Mesoscopic metal particles and wires by electrodeposition. *Journal of Physical Chemistrty B* **2002**, *106*, 3339–3353.

17. Perez, M. D., Otal, E., Bilmes, S. A., Soler-Illia, G., Crepaldi, E. L., Grosso, D., Sanchez, C., Growth of gold nanoparticle arrays in TiO_2 mesoporous matrixes. *Langmuir* **2004**, *20* (16), 6879–6886.

18. Rao, C. N. R., Ramakrishna Matte, H. S. S., Voggu, R., Govindaraj, A., Recent progress in the synthesis of inorganic nanoparticles. *Dalton Transactions* **2012**, *41* (17), 5089.

19. Sun, S., Zeng, H., Size-controlled synthesis of magnetite nanoparticles. *Journal of the American Chemical Society* **2002**, *124* (28), 8204–8205.

20. Sun, S., Zeng, H., Robinson, D. B., Raoux, S., Rice, P. M., Wang, S. X., Li, G., Monodisperse MFe_2O_4(M = Fe, Co, Mn) nanoparticles. *Journal of the American Chemical Society* 2004, *126* (1), 273–279.

21. Yin, Y., Alivisatos, A., Colloidal nanocrystal synthesis and the organic-inorganic interface. *Nature* **2005**, *437* (7059), 664–670.

22. Yong, K. T., Sahoo, Y., Choudhury, K. R., Swihart, M. T., Minter, J. R., Prasad, P. N., Shape control of PbSe nanocrystals using noble metal seed particles. *Nano Letters* **2006**, *6* (4), 709–714.

23. Yong, K. T., Sahoo, Y., Swihart, M. T., Prasad, P. N., Synthesis and plasmonic properties of silver and gold nanoshells on polystyrene cores of different size and of gold-silver core-shell nanostructures. *Colloid and Surface A: Physicochemical and Engineering Aspects* **2006**, *290* (1–3), 89–105.

24. Yong, K. T., Sahoo, Y., Swihart, M. T., Prasad, P. N., Growth of CdSe quantum rods and multipods seeded by noble-metal nanoparticles. *Advanced Materials* **2006**, *18* (15), 1978–1982.

25. Yong, K. T., Sahoo, Y., Swihart, M. T., Prasad, P. N., Shape control of CdS nanocrystals in one-pot synthesis. *Journal of Physical Chemistry C* **2007**, *111* (6), 2447–2458.

26. Yu, W. W., Falkner, J. C., Yavuz, C. T., Colvin, V. L., Synthesis of monodisperse iron oxide nanocrystals by thermal decomposition of iron carboxylate salts. *Chemical Communications* **2004**, (20), 2306–2307.

27. Yu, X., Chen, K., A facile surfactant-free fabrication of single-crystalline truncated Fe_3O_4 cubes. *Materials Science and Engineering B-Advanced Functional Solid-State Materials* **2011**, *176* (9), 750–755.

28. Tao, A. R., Habas, S., Yang, P. D., Shape control of colloidal metal nanocrystals. *Small* **2008**, *4* (3), 310–325.

29. Dobson, J., Cancer therapy: Death by magnetism. *Nature Publishing Group* **2012**, *11* (12), 1006–1008.

30. Fouriki, A., Dobson, J., Nanomagnetic gene transfection for non-viral gene delivery in NIH 3T3 mouse embryonic fibroblasts. *Materials* **2013**, *6* (1), 255–264.

31. Kozissnik, B., Bohórquez, A. C., Dobson, J., Rinaldi, C., Magnetic fluid hyperthermia: Advances, challenges, and opportunity. *International Journal of Hyperthermia* **2013**, *29* (8), 706–714.

32. Krishnan, K. M., Biomedical nanomagnetics: A spin through possibilities in imaging, diagnostics, and therapy. *IEEE Transactions on Magnetics* **2010**, *46* (7), 2523.

33. Torres-Diaz, I., Rinaldi, C., Recent progress in ferrofluids research: Novel applications of magnetically controllable and tunable fluids. *Soft Matter* **2014**, *10*, 8584–8602.

34. Jiles, D. C., Recent advances and future directions in magnetic materials. *Acta Materialia* **2003**, *51*, 5907–5939.

35. Naughton, B., Majewski, P., Clarke, D., Magnetic properties of nickel-zinc ferrite toroids prepared from nanoparticles. *Journal of the American Ceramic Society* **2007**, *90* (11), 3547–3553.

36. Wen, X., Starr, J. D., Andrew, J. S., Arnold, D., Electro-infiltration: A method to form nanocomposite soft magnetic cores for integrated magnetic devices. *Journal of Micromechanics and Microengineering* **2014**, *24*, 107001–107005.

37. Naughton, B. T., Clarke, D. R., Composition-size effects in nickel-zinc ferrite nanoparticles prepared by aqueous coprecipitation. *Journal of the American Ceramic Society* **2008**, *91* (4), 1253–1257.

38. Bee, A., Massart, R., Neveu, S., Synthesis of very fine maghemite particles. *Journal of Magnetism and Magnetic Materials* **1995**, *149*, 6–9.

39. Lefebure, S., Dubois, E., Cabuil, V., Neveau, S., Massart, R., Monodisperse magnetic nanoparticles: Preparation and dispersion in water and oils. *Journal of Materials Research* **1998**, *13* (10), 2975–2981.

40. Massart, R., Preparation of aqueous magnetic liquids in alkaline and acidic media. *IEEE Transactions on Magnetics* **1981**, *17*, 1247–1248.

41. Massart, R., Dubois, E., Cabuil, V., Hasmonay, E., Preparation and properties of monodisperse magnetic fluids. *Journal of Magnetism and Magnetic Materials* **1995**, *149*, 1–5.

42. Tourinho, F., Franck, R., Massart, R., Aqueous ferrofluids based on manganese and cobalt ferrites. *Journal of Materials Science* **1990**, *25*, 3249–3254.

43. Bronstein, L. M., Huang, X., Retrum, J., Schmucker, A., Pink, M., Stein, B. D., Dragnea, B., Influence of iron oleate complex structure on iron oxide nanoparticle formation. *Chemistry of Materials* **2007**, *19* (15), 3624–3632.

44. Calero, V. L., Gonzalez, A. M., Rinaldi, C., A statistical analysis to control the growth of cobalt ferrite nanoparticles synthesized by the thermodecomposition method. *Journal of Manufacturing Science and Engineering* **2010**, *132* (3), 030914.

45. Lam, C., Zhang, Y. F., Tang, Y. H., Lee, C. S., Bello, I., Lee, S. T., Large-scale synthesis of ultrafine Si nanoparticles by ball milling. *Journal of Crystal Growth* **2000**, *220* (4), 466–470.

46. Gajovic, A., Stubicar, M., Ivanda, M., Furic, K., Raman spectroscopy of ball-milled TiO_2. *Journal of Molecular Structure* **2001**, *563*, 315–320.

47. Aslibeiki, B., Kameli, P., Salamati, H., Eshraghi, M., Tahmasebi, T., Superspin glass state in $MnFe_2O_4$ nanoparticles. *Journal of Magnetism and Magnetic Materials* **2010**, *322* (19), 2929–2934.

48. Ferrari, M., Cancer nanotechnology: Opportunities and challenges. *Nature Reviews Cancer* **2005**, *5* (3), 161–171.

49. Rosi, N. L., Mirkin, C. A., Nanostructures in biodiagnostics. *Chemical Reviews* **2005**, *105* (4), 1547–1562.

50. Haynes, C. L., Van Duyne, R. P., Nanosphere lithography: A versatile nanofabrication tool for studies of size-dependent nanoparticle optics. *Journal of Physical Chemistry B* **2001**, *105* (24), 5599–5611.

51. Stewart, M. E., Anderton, C. R., Thompson, L. B., Maria, J., Gray, S. K., Rogers, J. A., Nuzzo, R. G., Nanostructured plasmonic sensors. *Chemical Reviews* **2008**, *108* (2), 494–521.

52. Jensen, T. R., Malinsky, M. D., Haynes, C. L., Van Duyne, R. P., Nanosphere lithography: Tunable localized surface plasmon resonance spectra of silver nanoparticles. *Journal of Physical Chemistry B* **2000**, *104* (45), 10549–10556.

53. Sardar, R., Funston, A. M., Mulvaney, P., Murray, R. W., Gold nanoparticles: Past, present, and future. *Langmuir* **2009**, *25* (24), 13840–13851.

54. Cheng, Y.-T., Liang, T., Nie, X., Choudhary, K., Phillpot, S. R., Asthagiri, A., Sinnott, S. B., Cu cluster deposition on ZnO(10–10): Morphology and growth mode predicted from molecular dynamics simulations. *Surface Science* **2014**, *621*, 109–116.

55. Wegner, K., Piseri, P., Tafreshi, H. V., Milani, P., Cluster beam deposition: A tool for nanoscale science and technology. *Journal of Physics D-Applied Physics* **2006**, *39* (22), R439–R459.

56. Erogbogbo, F., Yong, K. T., Roy, I., Xu, G. X., Prasad, P. N., Swihart, M. T., Biocompatible luminescent silicon quantum dots for imaging of cancer cells. *ACS Nano* **2008**, *2* (5), 873–878.

57. He, Y. Q., Sahoo, Y., Wang, S. M., Luo, H., Prasad, P. N., Swihart, M. T., Laser-driven synthesis and magnetic properties of iron nanoparticles. *Journal of Nanoparticle Research* **2006**, *8* (3–4), 335–342.

58. Smit, J., Wijn, H., *Ferrites: Physical Properties of Ferrimagnetic Oxides in Relation to Their Technical Applications.* New York, NY: John Wiley & Sons, **1959**.

59. LaMer, V. K., Dinegar, R. H., Theory, production and mechanism of formation of monodispersed hydrosols. *Journal of the American Chemical Society* **1950**, *72* (11), 4847–4854.

60. Murray, C. B., Kagan, C. R., Bawendi, M. G., Synthesis and characterization of monodisperse nanocrystals and close-packed nanocrystal assemblies. *Annual Review of Materials Science* **2000**, *30*, 545–610.

61. Sun, S. H., Anders, S., Thomson, T., Baglin, J. E. E., Toney, M. F., Hamann, H. F., Murray, C. B., Terris, B. D., Controlled synthesis and assembly of FePt nanoparticles. *Journal of Physical Chemistry B* **2003**, *107* (23), 5419–5425.

62. Greene, L. E., Yuhas, B. D., Law, M., Zitoun, D., Yang, P. D., Solution-grown zinc oxide nanowires. *Inorganic Chemistry* **2006**, *45* (19), 7535–7543.

63. Masala, O., Hoffman, D., Sundaram, N., Page, K., Proffen, T., Lawes, G., Seshadri, R., Preparation of magnetic spinel ferrite core/shell nanoparticles: Soft ferrites on hard ferrites and vice versa. *Solid State Sciences* **2006**, *8* (9), 1015–1022.

64. Masala, O., Spinel ferrite/MnO core/shell nanoparticles: Chemical synthesis of all-oxide exchange biased architectures. *Journal of the American Chemical Society* **2005**, *127*, 9354–9355.

65. Kwon, S. G., Piao, Y., Park, J., Angappane, S., Jo, Y., Hwang, N.-M., Park, J.-G., Hyeon, T., Kinetics of monodisperse iron oxide nanocrystal formation by "heating-up" process. *Journal of the American Chemical Society* **2007**, *129* (41), 12571–12584.

66. Na, H. B., Song, I. C., Hyeon, T., Inorganic nanoparticles for MRI contrast agents. *Advanced Materials* **2009**, *21*, 2133–2148.

67. Skrdla, P. J., Use of dispersive kinetic models for nucleation and denucleation to predict steady-state nanoparticle size distributions and the role of ostwald ripening. *Journal of Physical Chemistry C* **2012**, *116* (1), 214–225.

68. Gao, G., Liu, X., Shi, R., Zhou, K., Shi, Y., Ma, R., Takayama-Muromachi, E., Qiu, G., Shape-controlled synthesis and magnetic properties of monodisperse Fe_3O_4 nanocubes. *Crystal Growth Design* **2010**, *10* (7), 2888–2894.

69. Pankhurst, Q. A., Connolly, J., Jones, S. K., Dobson, J., Applications of magnetic nanoparticles in biomedicine. *Journal of Physics D: Applied Physics* **2003**, *36* (13), R167.

70. Pankhurst, Q. A., Thanh, N. T. K., Jones, S. K., Dobson, J., Progress in applications of magnetic nanoparticles in biomedicine. *Journal of Physics D: Applied Physics* **2009**, *42* (22), 224001.
71. Krishnan, K. M., Biomedical nanomagnetics: A spin through possibilities in imaging, diagnostics, and therapy. *IEEE Transactions on Magnetics* **2010**, *46* (7), 2523–2558.
72. Baek, M. J., Park, J. Y., Xu, W., Kattel, K., Kim, H. G., Lee, E. J., Patel, A. K., Lee, J. J., Chang, Y., Kim, T. J., Bae, J.-E., Chae, K.-S., Lee, G.-H., Water-soluble MnO nanocolloid for a molecular T-1 MR imaging: A facile one-pot synthesis, in vivo T-1 MR images, and account for relaxivities. *ACS Applied Materials Interfaces* **2010**, *2* (10), 2949–2955.
73. Guan, N., Wang, Y., Sun, D., Xu, J., A simple one-pot synthesis of single-crystalline magnetite hollow spheres from a single iron precursor. *Nanotechnology* **2009**, *20*, 105603.
74. Togashi, T., Naka, T., Asahina, S., Sato, K., Takami, S., Surfactant-assisted one-pot synthesis of superparamagnetic magnetite nanoparticle clusters with tunable cluster size and magnetic field sensitivity. *Dalton Transactions* **2010**, *40* (5), 1073–1078.
75. Lee, J. H., Jang, J. T., Choi, J. S., Moon, S. H., Noh, S. H., Kim, J. W., Kim, J. G., Kim, I. S., Park, K. I., Cheon, J., Exchange-coupled magnetic nanoparticles for efficient heat induction. *Nature Nanotechnology* **2011**, *6* (7), 418–422.
76. Laurent, S., Forge, D., Port, M., Roch, A., Robic, C., Elst, L. V., Muller, R. N., Magnetic iron oxide nanoparticles: Synthesis, stabilization, vectorization, physicochemical characterizations, and biological applications (2008, *108*, 2064). *Chem Rev* **2010**, *110* (4), 2574–2574.
77. Grun, M., Lauer, I., Unger, K. K., The synthesis of micrometer- and submicrometer-size spheres of ordered mesoporous oxide MCM-41. *Advanced Materials* **1997**, *9* (3), 254–257.
78. Lodha, A., Lodha, M., Patel, A., Chaudhuri, J., Dalal, J., Edwards, M., Douroumis, D., Synthesis of mesoporous silica nanoparticles and drug loading of poorly water soluble drug cyclosporin A. *Journal of Pharmacy and Bioallied Sciences* **2012**, *4* (Suppl 1), S92–S94.
79. Thanh, N. T. K., Green, L. A. W., Functionalisation of nanoparticles for biomedical applications. *Nano Today* **2010**, *5* (3), 213–230.
80. Hermanson, G. T., *Bioconjugate Techniques.* 2nd ed.; London, UK: Elsevier: 2008.
81. Chen, Z. P., Zhang, Y., Zhang, S., Xia, J. G., Liu, J. W., Xu, K., Gu, N., Preparation and characterization of water-soluble monodisperse magnetic iron oxide nanoparticles via surface double-exchange with DMSA. *Colloids and Surfaces A: Physicochemical and Engineering Aspects* **2008**, *316* (1–3), 210–216.
82. Liu, Y., Chen, Z., Wang, J., Systematic evaluation of biocompatibility of magnetic Fe3O4 nanoparticles with six different mammalian cell lines. *Journal of Nanoparticle Research* **2011**, *13* (1), 199–212.
83. Roca, A. G., Veintemillas-Verdaguer, S., Port, M., Robic, C., Serna, C. J., Morales, M. P., Effect of nanoparticle and aggregate size on the relaxometric properties of MR contrast agents based on high quality magnetite nanoparticles. *Journal of Physical Chemistry B* **2009**, *113* (19), 7033–7039.
84. Herranz, F., Morales, M. P., Roca, A. G., Desco, M., Ruiz-Cabello, J., A new method for the rapid synthesis of water stable superparamagnetic nanoparticles. *Chemistry—A European Journal* **2008**, *14* (30), 9126–9130.

85. Chen, D. X., Sun, N., Gu, H. C., Size analysis of carboxydextran coated superparamagnetic iron oxide particles used as contrast agents of magnetic resonance imaging. *Journal of Applied Physics* **2009**, *106*, 063906.
86. Chen, T.-J., Cheng, T.-H., Chen, C.-Y., Hsu, S. C. N., Cheng, T.-L., Liu, G.-C., Wang, Y.-M., Targeted Herceptin–dextran iron oxide nanoparticles for noninvasive imaging of HER2/neu receptors using MRI. *JBIC Journal of Biological Inorganic Chemistry* **2008**, *14* (2), 253–260.
87. DeNardo, S. J., DeNardo, G. L., Natarajan, A., Miers, L. A., Foreman, A. R., Gruettner, C., Adamson, G. N., Ivkov, R., Thermal dosimetry predictive of efficacy of In-111-ChL6 nanoparticle AMF-induced thermoablative therapy for human breast cancer in mice. *Journal of Nuclear Medicine* **2007**, *48* (3), 437–444.
88. Kozak, D., Chen, A., Bax, J., Trau, M., Protein resistance of dextran and dextran-poly(ethylene glycol) copolymer films. *Biofouling* **2011**, *27* (5), 497–503.
89. Krycka, K. L., Jackson, A. J., Borchers, J. A., Shih, J., Briber, R., Ivkov, R., Grüttner, C., Dennis, C. L., Internal magnetic structure of dextran coated magnetite nanoparticles in solution using small angle neutron scattering with polarization analysis. *Journal of Applied Physics* **2011**, *109* (7), 07B513.
90. Natarajan, A., Gruettner, C., Ivkov, R., DeNardo, G. L., Mirick, G., Yuan, A., Foreman, A., DeNardo, S. J., NanoFerrite particle based radioimmunonanoparticles: Binding affinity and in vivo pharmacokinetics. *Bioconjugate Chemistry* **2008**, *19* (6), 1211–1218.
91. Tassa, C., Shaw, S. Y., Weissleder, R., Dextran-coated iron oxide nanoparticles: A versatile platform for targeted molecular imaging, molecular diagnostics, and therapy. *Accounts of Chemical Research* **2011**, *44* (10), 842–852.
92. Zhang, J. L., Srivastava, R. S., Misra, R. D. K., Core-shell magnetite nanoparticles surface encapsulated with smart stimuli-responsive polymer: Synthesis, characterization, and LCST of viable drug-targeting delivery system. *Langmuir* **2007**, *23* (11), 6342–6351.
93. Creixell, M., Herrera, A. P., Latorre-Esteves, M., Ayala, V., Torres-Lugo, M., Rinaldi, C., The effect of grafting method on the colloidal stability and in vitro cytotoxicity of carboxymethyl dextran coated magnetic nanoparticles. *Journal of Material Chemistry* **2010**, *20* (39), 8539.
94. Herrera, A. P., Barrera, C., Rinaldi, C., Synthesis and functionalization of magnetite nanoparticles with aminopropylsilane and carboxymethyldextran. *Journal of Material Chemistry* **2008**, *18* (31), 3650.
95. Mefford, O., Woodward, R., Goff, J., Vadala, T., St Pierre, T., Dailey, J., Riffle, J., Field-induced motion of ferrofluids through immiscible viscous media: Testbed for restorative treatment of retinal detachment. *Journal of Magnetism and Magnetic Materials* **2007**, *311* (1), 347–353.
96. Mefford, O., Vadala, M., Goff, J., Carroll, M., Mejia-Ariza, R., Caba, B., Pierre, T., Woodward, R., Davis, R., Riffle, J., Stability of polydimethylsiloxane-magnetite nanoparticle dispersions against flocculation: Interparticle interactions of polydisperse materials. *Langmuir* **2008**, *24* (9), 5060–5069.
97. Mefford, O. T., Carroll, M. R. J., Vadala, M. L., Goff, J. D., Mejia-Ariza, R., Saunders, M., Woodward, R. C., St Pierre, T. G., Davis, R. M., Riffle, J. S., Size analysis of PDMS-magnetite nanoparticle complexes: Experiment and theory. *Chemisty of Materials* **2008**, *20* (6), 2184–2191.

98. Miles, W. C., Goff, J. D., Huffstetler, P. P., Mefford, O. T., Riffle, J. S., Davis, R. M., The design of well-defined PDMS-Magnetite complexes. *Polymer* 2010, *51* (2), 482–491.
99. Herrera, A. P., Rodríguez, M., Torres-Lugo, M., Rinaldi, C., Multifunctional magnetite nanoparticles coated with fluorescent thermo-responsive polymeric shells. *Journal of Material Chemistry* 2008, *18* (8), 855.
100. Isojima, T., Lattuada, M., Vander Sande, J. B., Hatton, T. A., Reversible clustering of pH- and temperature-responsive janus magnetic nanoparticles. *ACS Nano* 2008, *2* (9), 1799–1806.
101. Purushotham, S., Chang, P., Rumpel, H., Kee, I., Ng, R., Chow, P., Tan, C., Ramanujan, R., Thermoresponsive core–shell magnetic nanoparticles for combined modalities of cancer therapy. *Nanotechnology* 2009, *20*, 305101.
102. Zhou, L., Yuan, J., Core–shell structural iron oxide hybrid nanoparticles: From controlled synthesis to biomedical applications. *Journal of Materials Chemistry* 2010.
103. Lien, Y.-H., Wu, T.-M., Wu, J.-H., Liao, J.-W., Cytotoxicity and drug release behavior of PNIPAM grafted on silica-coated iron oxide nanoparticles. *Journal of Nanoparticle Research* 2011, *13* (10), 5065–5075.
104. Mincheva, R., Stoilova, O., Penchev, H., Ruskov, T., Spirov, I., Manolova, N., Rashkov, I., Synthesis of polymer-stabilized magnetic nanoparticles and fabrication of nanocomposite fibers thereof using electrospinning. *European Polymer Journal* 2008, *44* (3), 615–627.
105. Bae, K. H., Park, M., Do, M. J., Lee, N., Ryu, J. H., Kim, G. W., Kim, C., Park, T. G., Hyeon, T., Chitosan oligosaccharide-stabilized ferrimagnetic iron oxide nanocubes for magnetically modulated cancer hyperthermia. *ACS Nano* 2012, *6* (6), 5266–5273.
106. Bhattacharya, D., Das, M., Mishra, D., Banerjee, I., Sahu, S. K., Maiti, T. K., Pramanik, P., Folate receptor targeted, carboxymethyl chitosan functionalized iron oxide nanoparticles: A novel ultradispersed nanoconjugates for bimodal imaging. *Nanoscale* 2011, *3* (4), 1653.
107. López-Cruz, A., Barrera, C., Calero-DdelC, V. L., Rinaldi, C., Water dispersible iron oxide nanoparticles coated with covalently linked chitosan. *Journal of Materials Chemistry* 2009, *19* (37), 6870.
108. Zhao, D.-L., Wang, X.-X., Zeng, X.-W., Xia, Q.-S., Tang, J.-T., Preparation and inductive heating property of Fe3O4–chitosan composite nanoparticles in an AC magnetic field for localized hyperthermia. *Journal of Alloys and Compounds* 2009, *477* (1–2), 739–743.
109. Townsend, J., Burtovyy, R., Galabura, Y., Luzinov, I., Flexible chains of ferromagnetic nanoparticles. *ACS Nano* 2014, 8 (7), 6970–6978.
110. Sun, C., Sze, R., Zhang, M., Folic acid-PEG conjugated superparamagnetic nanoparticles for targeted cellular uptake and detection by MRI. *Journal of Biomedical Materials Research Part A* 2006, *78A* (3), 550–557.
111. Chen, S., Li, Y., Guo, C., Wang, J., Ma, J., Liang, X., Yang, L.-R., Liu, H.-Z., Temperature-responsive magnetite/PEO–PPO–PEO block copolymer nanoparticles for controlled drug targeting delivery. *Langmuir* 2007, *23* (25), 12669–12676.
112. Aqil, A., Vasseur, S., Duguet, E., Passirani, C., Benoît, J. P., Roch, A., Muller, R., Jérôme, R., Jérôme, C., PEO coated magnetic nanoparticles for biomedical application. *European Polymer Journal* 2008, *44* (10), 3191–3199.

113. Feng, B., Hong, R. Y., Wang, L. S., Guo, L., Li, H. Z., Ding, J., Zheng, Y., Wei, D. G., Synthesis of Fe3O4/APTES/PEG diacid functionalized magnetic nanoparticles for MR imaging. *Colloids and Surfaces A: Physicochemical and Engineering Aspects* **2008**, *328* (1–3), 52–59.
114. Yallapu, M. M., Foy, S. P., Jain, T. K.; Labhasetwar, V., PEG-functionalized magnetic nanoparticles for drug delivery and magnetic resonance imaging applications. *Pharmaceutical Research* **2010**, *27* (11), 2283–2295.
115. Liu, D., Wu, W., Ling, J., Wen, S., Gu, N., Zhang, X., Effective PEGylation of iron oxide nanoparticles for high performance in vivo cancer imaging. *Advanced Functional Materials* **2011**, *21* (8), 1498–1504.
116. Viali, W. R., Silva Nunes, E., Santos, C. C., Silva, S. W., Aragón, F. H., Coaquira, J. A. H., Morais, P. C., Jafelicci, M., PEGylation of SPIONs by polycondensation reactions: A new strategy to improve colloidal stability in biological media. *Journal of Nanoparticle Research* **2013**, *15* (8), 1824.
117. Stone, R. C., Qi, B., Trebatoski, D., Jetti, R., Bandera, Y. P., Foulger, S. H., Mefford, O. T., A versatile stable platform for multifunctional applications: Synthesis of a nitroDOPA–PEO–alkyne scaffold for iron oxide nanoparticles. *Journal of Materials Chemistry B* **2014**, *2* (30), 4789.
118. Barrera, C., Herrera, A. P., Bezares, N., Fachini, E., Olayo-Valles, R., Hinestroza, J. P., Rinaldi, C., Effect of poly(ethylene oxide)-silane graft molecular weight on the colloidal properties of iron oxide nanoparticles for biomedical applications. *Journal of Colloid and Interface Science* **2012**, *377* (1), 40–50.
119. Barrera, C., Herrera, A. P., Rinaldi, C., Colloidal dispersions of monodisperse magnetite nanoparticles modified with poly(ethylene glycol). *Journal of Colloid and Interface Science* **2009**, *329* (1), 107–113.
120. Reimer, P., Balzer, T., Ferucarbotran (Resovist): A new clinically approved RES-specific contrast agent for contrast-enhanced MRI of the liver: Properties, clinical development, and applications. *European Radiology* **2003**, *13* (6), 1266–1276.
121. Laniado, M., Chachuat, A., [The endorem tolerance profile]. *Der Radiologe* **1995**, *35* (11 Suppl 2), S266–70.
122. Mao, H. Q., Roy, K., Troung-Le, V. L., Janes, K. A., Lin, K. Y., Wang, Y., August, J. T., Leong, K. W., Chitosan-DNA nanoparticles as gene carriers: Synthesis, characterization and transfection efficiency. *Journal of Controlled Release* **2001**, *70* (3), 399–421.
123. Yang, S. J., Chang, S. M., Tsai, K. C., Tsai, H. M., Chen, W. S., Shieh, M. J., Enhancement of chitosan nanoparticle-facilitated gene transfection by ultrasound both in vitro and in vivo. *Journal of Biomedical Materials Research B* **2012**, *100B* (7), 1746–1754.
124. Sun, Y., Zhang, S., Peng, X., Gong, Z. Y., Li, X. L., Yuan, Z. Q., Li, Y., Zhang, D. W., Peng, Y. Z., Preparation, characterization and transfection efficacy of chitosan nanoparticles containing the intestinal trefoil factor gene. *Molecular Biology Reports* **2012**, *39* (2), 945–952.
125. Vadala, M. L., Thompson, M. S., Ashworth, M. A., Lin, Y., Vadala, T. P., Ragheb, R., Riffle, J. S., Heterobifunctional poly(ethylene oxide) oligomers containing carboxylic acids. *Biomacromolecules* **2008**, *9* (3), 1035–1043.
126. Cammas, S., Nagasaki, Y., Kataoka, K., Heterobifunctional Poly (ethylene oxide): Synthesis of alpha-Methoxy. omega-amino and alpha-Hydroxy- omega-amino PEOs with the Same Molecular Weights. *Bioconjugate Chemistry* **1995**, *6* (2), 226–230.

127. Amstad, E., Zurcher, S., Mashaghi, A., Wong, J. Y., Textor, M., Reimhult, E., Surface functionalization of single superparamagnetic iron oxide nanoparticles for targeted magnetic resonance imaging. *Small* **2009**, *5* (11), 1334–42.
128. Kim, J., Lee, J. E., Lee, S. H., Yu, J. H., Lee, J. H., Park, T. G., Hyeon, T., Designed fabrication of a multifunctional polymer nanomedical platform for simultaneous cancer-targeted imaging and magnetically guided drug delivery. *Advanced Materials* **2008**, *20* (3), 478–483.
129. Harris, J. M., Chess, R. B., Effect of pegylation on pharmaceuticals. *Nature Reviews. Drug Discovery.* **2003**, *2* (3), 214–221.
130. Zhang, Y., Kohler, N., Zhang, M., Surface modification of superparamagnetic magnetite nanoparticles and their intracellular uptake. Biomaterials **2002**, *23* (7), 1553–1561.
131. Thompson, M., Vadala, T., Vadala, M., Lin, Y., Riffle, J., Synthesis and applications of heterobifunctional poly (ethylene oxide) oligomers. *Polymer* **2007**, *49*, 345–373.
132. Crist, R. M., Grossman, J. H., Patri, A. K., Stern, S. T., Dobrovolskaia, M. A., Adiseshaiah, P. P., Clogston, J. D., McNeil, S. E., Common pitfalls in nanotechnology: Lessons learned from NCI's Nanotechnology Characterization Laboratory. *Integrative Biology : Quantitative Biosciences from Nano to Macro* **2013**, *5* (1), 66–73.
133. Kozissnik, B., Green, L. A. W., Chester, K. A., Thanh, N. T. K., Strategy for functionalisation of MNPs for specific targets. In *Magnetic Nanoparticles: From Fabrication to Clinical Applications*, Thanh, N. T. K., Ed. CRC Press, Boca Raton, FL: **2012**, pp 127–150.
134. Amstad, E., Textor, M., Reimhult, E., Stabilization and functionalization of iron oxide nanoparticles for biomedical applications. *Nanoscale* **2011**, *3* (7), 2819–2843.
135. Goff, J. D., Huffstetler, P. P., Miles, W. C., Pothayee, N., Reinholz, C. M., Ball, S., Davis, R. M., Riffle, J. S., Novel Phosphonate-functional poly(ethylene oxide)-magnetite nanoparticles form stable colloidal dispersions in phosphate-buffered saline. *Chemistry of Materials* **2009**, *21* (20), 4784–4795.
136. Davis, K. M., Qi, B., Witmer, M., Kitchens, C. L., Powell, B. A., Mefford, O. T., Quantitative measurement of ligand exchange on iron oxide nanoparticles via radiolabeled oleic acid. *Langmuir* **2014**, *30* (36), 10918–10925.
137. Amstad, E., Gehring, A. U., Fischer, H., Nagaiyanallur, V. V., Hahner, G., Textor, M., Reimhult, E., Influence of electronegative substituents on the binding affinity of catechol-derived anchors to Fe3O4 nanoparticles. *Journal of Physical Chemistry C* **2011**, *115* (3), 683–691.
138. Saville, S. L., Stone, R. C., Qi, B., Mefford, O. T., Investigation of the stability of magnetite nanoparticles functionalized with catechol based ligands in biological media. *Journal of Materials Chemistry* **2012**, *22*, 24909.
139. Perumal, S., Hofmann, A., Scholz, N., Rühl, E., Graf, C., Kinetics study of the binding of multivalent ligands on size-selected gold nanoparticles. *Langmuir* **2011**, *27* (8), 4456–4464.
140. Na, H. B., Palui, G., Rosenberg, J. T., Ji, X., Grant, S. C., Mattoussi, H., Multidentate catechol-based polyethylene glycol oligomers provide enhanced stability and biocompatibility to iron oxide nanoparticles. *ACS Nano* **2012**, *6* (1), 389–399.
141. Ujiie, K., Kanayama, N., Asai, K., Kishimoto, M., Ohara, Y., Akashi, Y., Yamada, K., Hashimoto, S., Oda, T., Ohkohchi, N., Yanagihara, H., Kita, E., Yamaguchi, M., Fujii, H., Nagasaki, Y., Preparation of highly dispersible and

tumor-accumulative, iron oxide nanoparticles: Multi-point anchoring of PEG-b-poly(4-vinylbenzylphosphonate) improves performance significantly. *Colloids and Surfaces B: Biointerfaces* **2011**, *88* (2), 771–778.

142. Cruz-Acuña, M., Maldonado-Camargo, L., Dobson, J., Rinaldi, C., From oleic acid-capped iron oxide nanoparticles to polyethyleneimine-coated single-particle magnetofectins. *Journal of Nanoparticle Research* **2016**, *18* (9), 268.

143. Duan, H., Kuang, M., Wang, X., Wang, Y. A., Mao, H., Nie, S., Reexamining the effects of particle size and surface chemistry on the magnetic properties of iron oxide nanocrystals: New insights into spin disorder and proton relaxivity. *Journal of Physical Chemistry C* **2008**, *112* (22), 8127–8131.

144. Bishop, K. J. M., Wilmer, C. E., Soh, S., Grzybowski, B. A., Nanoscale forces and their uses in self-assembly. *Small* **2009**, *5* (14), 1600–1630.

145. Hiemenz, P. C., Rajagopalan, R., *Principles of Colloid and Surface Chemistry*. Boca Raton, FL: Taylor & Francis: **1997**.

146. Pecora, R., Dynamic light scattering measurement of nanometer particles in liquids. *Journal of Nanoparticle Research* **2000**, *2* (2), 123–131.

147. Berg, J. C., *An Introduction to Interfaces and Colloids: The Bridge to Nanoscience*. Singapore: World Scientific Publishing: **2010**.

148. Nappini, S., Bonini, M., Bombelli, F., Pineider, F., Controlled drug release under a low frequency magnetic field: Effect of the citrate coating on magnetoliposomes stability. *Soft Matter* **2010**, *7*, 1025–1037.

149. Santiago-Quinones, D., Acevedo, A., Rinaldi, C., Magnetic and magnetorheological characterization of a polymer liquid crystal ferronematic. *Journal of Applied Physics* **2009**, *105* (7), 07B512.

150. Butter, K., Bomans, P., Frederik, P. M., Vroege, G. J., Philipse, A. P., Direct observation of dipolar chains in iron ferrofluids by cryogenic electron microscopy. *Nature Materials* **2003**, *2* (2), 88–91.

151. Balasubramaniam, S., Kayandan, S., Lin, Y.-N., Kelly, D. F., House, M. J., Woodward, R. C., St Pierre, T. G., Riffle, J. S., Davis, R. M., Toward design of magnetic nanoparticle clusters stabilized by biocompatible diblock copolymers for T 2-weighted MRI contrast. *Langmuir* **2014**, *30* (6), 1580–1587.

152. Bonini, M., Lenz, S., Falletta, E., Ridi, F., Acrylamide-based magnetic nanosponges: A new smart nanocomposite material. *Langmuir* **2008**, *24* (21), 12644–12650.

153. Bonini, M., Lenz, S., Giorgi, R., Baglioni, P., Nanomagnetic sponges for the cleaning of works of art. *Langmuir* **2007**, *23* (17), 8681–8685.

154. Farrell, D., Ijiri, Y., Kelly, C., Borchers, J., Rhyne, J., Ding, Y., Majetich, S., Small angle neutron scattering study of disordered and crystalline iron nanoparticle assemblies. *Journal of Magnetism and Magnetic Materials* **2006**, *303* (2), 318–322.

155. Klokkenburg, M., Erné, B. H., Wiedenmann, A., Petukhov, A. V., Philipse, A. P., Dipolar structures in magnetite ferrofluids studied with small-angle neutron scattering with and without applied magnetic field. *Physical Review E* **2007**, *75* (5), 51408.

156. Holm, C., Weis, J., The structure of ferrofluids: A status report. *Current Opinion in Colloid & Interface Science* **2005**, *10* (3–4), 133–140.

157. Carroll, M., Woodward, R., House, M., Teoh, W., Amal, R., Hanley, T., St .Pierre, T., Experimental validation of proton transverse relaxivity models for superparamagnetic nanoparticle MRI contrast agents. *Nanotechnology* **2010**, *21*, 035103.

158. Amstad, E., Kohlbrecher, J., Müller, E., Schweizer, T., Textor, M., Reimhult, E., Triggered release from liposomes through magnetic actuation of iron oxide nanoparticle containing membranes. *Nano Letters* **2011**, 5544–5549.
159. Dennis, C. L., Jackson, A. J., Borchers, J. A., Hoopes, P. J., Strawbridge, R., Foreman, A. R., van Lierop, J., Grüttner, C., Ivkov, R., Nearly complete regression of tumors via collective behavior of magnetic nanoparticles in hyperthermia. *Nanotechnology* **2009**, *20*, 395103.
160. Herrera, A. P., Barrera, C., Zayas, Y., Rinaldi, C., Monitoring colloidal stability of polymer-coated magnetic nanoparticles using AC susceptibility measurements. *Journal of Colloid and Interface Science* **2010**, *342* (2), 540–549.
161. Krycka, K. L., Booth, R. A., Hogg, C. R., Ijiri, Y., Borchers, J. A., Chen, W. C., Watson, S. M., Laver, M., Gentile, T. R., Dedon, L. R., Harris, S., Rhyne, J. J., Majetich, S. A., Core-shell magnetic morphology of structurally uniform magnetite nanoparticles. *Physical Review Letters* **2010**, *104* (20), 207203.
162. Grüttner, C., Müller, K., Teller, J., Westphal, F., Synthesis and functionalisation of magnetic nanoparticles for hyperthermia applications. *International Journal of Hyperthermia* **2013**, *29* (8), 777–789.
163. Veiseh, O., Gunn, J. W., Zhang, M., Design and fabrication of magnetic nanoparticles for targeted drug delivery and imaging. *Advanced Drug Delivery Reviews* **2010**, *62* (3), 284–304.
164. Tan, H., Yu, L., Gao, F., Liao, W., Wang, W., Zeng, W., Surface modification: How nanoparticles assemble to molecular imaging probes. *Journal of Nanoparticle Research* **2013**, *15* (12), 2100.
165. Niemeyer, C. M., *Bioconjugation Protocols: Strategies and Methods.* New York, NY: Springer Science & Business Media: **2004**; Vol. 283.
166. Kohler, N., Fryxell, G. E., Zhang, M., A bifunctional poly (ethylene glycol) silane immobilized on metallic oxide-based nanoparticles for conjugation with cell targeting agents. *Journal of the American Chemical Society* **2004**, *126* (23), 7206–7211.
167. Wang, A. Z., Bagalkot, V., Vasilliou, C. C., Gu, F., Alexis, F., Zhang, L., Shaikh, M., Yuet, K., Cima, M. J., Langer, R., Superparamagnetic iron oxide nanoparticle–aptamer bioconjugates for combined prostate cancer imaging and therapy. *ChemMedChem* **2008**, *3* (9), 1311–1315.
168. Padwa, A., *1, 3-Dipolar Cycloaddition Chemistry.* New York, NY: Wiley-Interscience: **1984**; Vol. 1.
169. Rostovtsev, V. V., Green, L. G., Fokin, V. V., Sharpless, K. B., A stepwise huisgen cycloaddition process: Copper (I)-catalyzed regioselective "ligation" of azides and terminal alkynes. *Angewandte Chemie International Edition* **2002**, *114* (14), 2708–2711.
170. Hein, C. D., Liu, X.-M., Wang, D., Click chemistry, a powerful tool for pharmaceutical sciences. *Pharmaceutical Research* **2008**, *25* (10), 2216–2230.
171. Seo, T. S., Li, Z., Ruparel, H., Ju, J., Click chemistry to construct fluorescent oligonucleotides for DNA sequencing. *Journal of Organic Chemistry* **2003**, *68* (2), 609–612.
172. Bolley, J., Guenin, E., Lievre, N., Lecouvey, M., Soussan, M., Lalatonne, Y., Motte, L., Carbodiimide versus click chemistry for nanoparticle surface functionalization: A comparative study for the elaboration of multimodal superparamagnetic nanoparticles targeting α vβ 3Integrins. *Langmuir* **2013**, *29* (47), 14639–14647.

173. Laughlin, S. T., Baskin, J. M., Amacher, S. L., Bertozzi, C. R., In vivo imaging of membrane-associated glycans in developing zebrafish. *Science* **2008**, *320* (5876), 664–667.
174. Ning, X., Temming, R. P., Dommerholt, J., Guo, J., Ania, D. B., Debets, M. F., Wolfert, M. A., Boons, G.-J., van Delft, F. L., Protein modification by strain-poromoted alkyne-nitrone cycloaddition. *Angewandte Chemie International Edition* **2010**, *49* (17), 3065–3068.
175. Hoyle, C. E., Bowman, C. N., Thiol-ene click chemistry. *Angewandte Chemie International Edition* **2010**, *49* (9), 1540–1573.
176. van der Vlies, A. J., O'Neil, C. P., Hasegawa, U., Hammond, N., Hubbell, J. A., Synthesis of pyridyl disulfide-functionalized nanoparticles for conjugating thiol-containing small molecules, peptides, and proteins. *Bioconjugate Chemistry* **2010**, *21* (4), 653–662.

3

Thermomagnetic Activation of Cellular Ion Channels to Control Cellular Activity

Heng Huang and Arnd Pralle

CONTENTS

3.1 Introduction 41
    3.1.1 Cell Signaling via Ion Channels 42
    3.1.2 Thermomagnetic Cell Stimulation: Coupling a
        Temperature-Sensitive Ion Channel with
        Local NP-Based Heating 43
3.2 Details of Local Magnetic NP-Based Hyperthermia 44
    3.2.1 Power Dissipation in a Magnetic NP Suspension 45
    3.2.2 Relaxation Mechanisms 46
    3.2.3 Heat Transfer from NP to Thermosensitive Protein 47
3.3 Experimental Considerations for Thermomagnetic Stimulation 48
    3.3.1 NP Synthesis and Functionalization 48
    3.3.2 NP Physical and Magnetic Characterization 49
    3.3.3 NP Heating Capacity: SAR 50
    3.3.4 Measuring Local Heating Using Ence Quantum
        Yield and Lifetime 51
    3.3.5 Local Temperature on NP Surface and in the Cell Membrane 53
3.4 Applications 53
    3.4.1 Remote Triggering of Cellular Calcium Influx 53
    3.4.2 Remote Triggering Action Potentials and Behavior 54
    3.4.3 Remote Activation of Protein Production and Secretion 55
3.5 Summary and Conclusions 56
References 57

3.1 Introduction

Cells, the basic building blocks of life, are spatially structured, yet highly dynamic factories that produce almost everything an organism needs. Their functions require that they communicate with each other on many levels and timescales, from millisecond electronic communication in the heart to synchronize contraction, to communication lasting minutes, hours, and even days, and over long distances across the body via chemical messengers, such as hormones. Many human diseases are caused by erroneous signal procession: diabetes is either caused by the inability to produce the signaling hormone insulin or by the desensitization of its receptor; the majority of cancers result from erroneous processing of growth signaling delivered by growth hormones. Hence, the development of tools to communicate with cells inside the intact organism is of great interest for basic research and new therapeutic approaches. Ideally, we find ways to read the state of single cell and then relate back a message to a cell without having to disrupt the organism.

Over the last quarter of century, optical methods have revolutionized biomedical research and knowledge by not only imaging structures at increasingly higher temporal and spatial resolution, but especially by measuring functional states of molecules inside cells and organisms. However, until recently there was no way to send a message to a cell other than by adding a diffusive ligand or poking an electrode into the cell. Both methods have severe technical limitation: the former is imprecise, the later very invasive. In 2005, optical stimulation methods, combining genetics and light to specifically activate or silence cells, were developed.[1-3] These have begun to revolutionize the analysis of neuronal circuitries and other cellular communication networks. However, the visible light used in this approach penetrates tissue poorly requiring piercing an optical fiber into the brain.[4]

Hence, an alternative truly remote method is desirable. Magnetic fields penetrate tissue much more easily and magnetic stimulation would provide a series of benefits: (1) nearly unlimited penetration depth in tissue, (2) parallel stimulation of multiple freely moving animals, and (3) compatible with optical stimulation and measurement methods. Development of *in vivo* magnetic cell stimulation is hindered because the weak interaction between magnetic fields and biomaterial require a transducer to translate the magnetic signal into a biological stimulus. In the past, magnetic nanoparticles (NPs) or microparticles were used to exert force, torque, or to aggregate receptors.[5,6] Here, we review how coupling local heating of superparamagnetic NPs in a radio-frequency (RF) magnetic field with thermosensitive ion channels may be used to remotely stimulate cells in- and ex vivo. This chapter describes method details, such as NP selection, coating, and delivery, and shows early applications, for example, remote neurostimulation and remote activation of protein production *in vivo*.

3.1.1 Cell Signaling via Ion Channels

Cell signaling is required in finding food or detecting danger, and in multicellular organisms it provides the basis for concerted function required for organs. Ion channels are one class of molecules used for sensing and signaling. They function as preassembled protein dimer or tetramer with typically one or two pores to conduct ions. The conduction pore is either cation or anion selective, and often specific for just one type of ions K^+,

Na$^+$, Ca^{2+}, or Cl$^-$. The conduction pore is usually closed by a gate, which opens in response to ligand binding or a change of a physical parameter, such as the voltage across the membrane, membrane tension, or temperature. Especially membrane voltage is used in organs requiring fast signal processing and spread such as the nervous system and all muscles. Ion channels gate within milliseconds and the flow of ions through pores in the membrane influences the membrane potential directly. The membrane potential of excitatory cells, such as neurons and muscle cells, is at rest negative, making these cells polarized. When cations flow inward, these cells are depolarized. Neurons contain voltage-gated channels, which provide a positive feedback to accelerate the depolarization into a signal spike. Therefore, opening just a few cation channels may activate these cells. The membrane potential is set mostly by the concentration of K$^+$, Na$^+$, and Cl$^-$ ions. Calcium is an important messenger molecule, which activates or deactivates many proteins by direct binding. Hence, opening calcium channels may cause muscle contraction, activate other channels, or activate gene transcription. Recently, calcium channels were successfully activated in RF magnetic fields combining nanomagnets as local heaters with a temperature-sensitive cation channel. This approach was shown in one study to produce sufficient amount of calcium influx to activate action potentials in neurons and in another study to upregulate transcription. Here, we review the essential requirements for successful cell stimulation by local heating of superparamagnetic NPs in RF magnetic fields.

3.1.2 Thermomagnetic Cell Stimulation: Coupling a Temperature-Sensitive Ion Channel with Local NP-Based Heating

Figure 3.1 shows the principle of how local heating of superparamagnetic in an RF magnetic field heats and opens a temperature-sensitive ion channel, which subsequently become conductive for cations, Na$^+$ and Ca^{2+}.[7] The influx of these cations may then either signal directly to the cell by depolarizing the electrical potential across the cell membrane, and in neurons trigger action potential firing, or it may evoke a secondary response through binding to regulator proteins inside the cell, which may initiate protein expression.

The TRPV1 channel is well suited for thermomagnetic stimulation because it is activated around 42°C,[8,9] which is sufficiently near normal body temperature to permit quick stimulation while allowing the channels to be normally closed. TRPV1 has been heterologously expressed in *Drosophila* neurons and stimulated with capsaicin to successfully evoke behavioral responses.[10] In addition, the TRP channel family contains a large variety of ion channels with different temperature sensitivities and conductances, which will allow expanding the thermomagnetic stimulation toolbox. Furthermore, thermosensitive anion channels may be used to hyperpolarize and silence neurons.

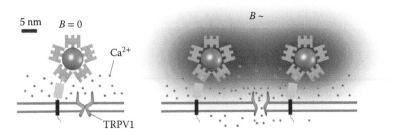

FIGURE 3.1

Principle of ion-channel stimulation using nanoparticle (NP) heating in an alternating external magnetic field. Streptavidin–DyLight549 (orange)-coated manganese ferrite ($MnFe_2O_4$) NPs ($d = 6$ nm, gray dots) in an RF magnetic field (B~) heats (red) and opens temperature-sensitive ion channel, TRPV1. The NPs are targeted to the cell and the vicinity by binding to the AP-CFP-TM protein through the biotinylated AP domain (green box), which is anchored to the membrane by the TM (blue box) and CFP (cyan box) domains. (Adapted from Huang, H. et al., *Nat Nanotechnol*, 5, 602–6, 2010.)

Other temperature-sensitive proteins may be engineered for various signaling modalities.

The cell-specific targeting of the stimulation within the body is a result of specific NP binding to the cells also expressing the temperature-sensitive channel. The magnetic field does not need to be focused. Indeed, the method activates all cells with NPs and TPRV1 channel uniformly across a large volume, making it feasible for *in vivo* whole body applications. The NPs should always be specifically bound to the target cells via an antibody–antigen or similar stabile bound to provide sufficient specificity. Figure 3.1 shows targeting via streptavidin-biotin linkage to an anchor protein separate from the ion channel. This approach provides separate control over the expression level of the channel and the number of NPs. Alternatively, the NPs may be targeted directly to the TRPV1 channel by engineering an epitope tag into an extracellular loop of the TRPV1 channel and then using antibodies to bind the NP.[11]

The main challenge for any remote activation of proteins or cells is to deliver sufficient energy for fast activation without causing tissue damage. In the case of the thermomagnetic approach, the rate-limiting factor is the speed of local heating, which in early versions took several seconds, but may be reduced.

3.2 Details of Local Magnetic NP-Based Hyperthermia

Prerequisites for effective remote control of cellular signaling are efficient local heating. The theory of superparamagnetic NP hyperthermia describes the dependence of NP heating on external field strength, frequency and NP material properties and size. In addition, the local particle density and geometric arrangement affect the heat delivery to temperature-sensitive protein.

3.2.1 Power Dissipation in a Magnetic NP Suspension

A theoretical description of power dissipation in magnetic fluid subjected to alternating magnetic field is provided by Rosensweig.[12] For a magnetic fluid with constant density in adiabatic conditions, the change in its internal energy, U, is equal to the work done by the magnetic field, therefore

$$dU = H \cdot dB \tag{3.1}$$

in which H is the magnetic field intensity and B the magnetic induction in the sample. H and B are in the same direction, therefore $dU = H \cdot dB$. The magnitude of the induction can be written as

$$B = \mu_0 (H + M) \tag{3.2}$$

where μ_0 is the permeability of free space and M the magnetization. Substituting Equation 3.2 into Equation 3.1 and integrating by parts over a cycle gives the cyclic increase of internal energy

$$\Delta U = -\mu_0 \oint M \, dH \tag{3.3}$$

The alternating magnetic field can be expressed as

$$H(t) = H_0 \cos \omega t = \text{Re}\left[H_0 e^{i\omega t}\right] \tag{3.4}$$

And M can be expressed in terms of the complex susceptibility $\chi = \chi' - i\chi''$, yielding

$$M(t) = \text{Re}\left[\chi H_0 e^{i\omega t}\right] = H_0 \left(\chi' \cos \omega t + \chi'' \sin \omega t\right) \tag{3.5}$$

Substituting Equations 3.4 and 3.5 to Equation 3.3 leads to

$$\Delta U = 2\mu_0 H_0^2 \chi'' \int_0^{2\pi/\omega} \sin^2 \omega t \, dt \tag{3.6}$$

Integrating and multiplying the results by the frequency $f = \omega/2\pi$ gives the expression for volumetric power dissipation:

$$P = \mu_0 \pi \chi'' f H_0^2 \tag{3.7}$$

The imaginary part of susceptibility, χ'', is dependent on both material properties and external field frequency:

$$\chi'' = \frac{\omega\tau}{1+(\omega\tau)^2}\chi_0 \tag{3.8}$$

where τ is the relaxation time. Substituting the above expression of χ'' into Equation 3.7 yields power dissipation in terms of magnetic field oscillation frequency f, field intensity H_0, and relaxation time τ:

$$P = \mu_0\pi f H_0{}^2\chi_0 \frac{2\pi f\tau}{1+(2\pi f\tau)^2} \tag{3.9}$$

It is clear that the power loss is proportional to $H_0{}^2$. At low frequencies, P is proportional to f^2 while it saturates at higher frequencies.

3.2.2 Relaxation Mechanisms

For frequencies below the ferromagnetic resonance frequency, Brownian and Néel relaxation are the two main physical mechanisms for power loss.[12,13] Brownian relaxation is due to the physical rotation of the particle. The relaxation time is given by

$$\tau_B = \frac{3\eta V_H}{kT} \tag{3.10}$$

in which η is the viscosity of the medium, and V_H the hydrodynamic volume of the particle.

Néel relaxation is due to the rotation of the magnetization, and its relaxation time is given by

$$\tau_N = \frac{\sqrt{\pi}}{2}\tau_0\frac{e^{\Gamma}}{\Gamma^{\frac{1}{2}}}, \qquad \Gamma = \frac{KV_M}{kT} \tag{3.11}$$

where K is the anisotropy constant, V_M the magnetic volume, and τ_0 has a typical value of 10^{-9} seconds. The two mechanisms operate in parallel, giving the expression for the overall relaxation time

$$\frac{1}{\tau} = \frac{1}{\tau_B} + \frac{1}{\tau_N} \tag{3.12}$$

In most biological applications, the hydrodynamic diameter of the NPs is so large that Brownian relaxation does not contribute significantly to the heating. In addition, the tethering of the NPs to the membrane required for cell stimulation strongly limits their rotation freedom.

3.2.3 Heat Transfer from NP to Thermosensitive Protein

An isolated NP delivers only a few femtowatts of power and theoretically the energy would dissipate too quickly into the surrounding water to affect nearby proteins. On cells, the NP can be sufficiently densely spaced in two dimensions so that their lateral heat field overlap and the cell membrane is heated efficiently without heating the inside of the cells (Figure 3.2).

However, the details of the heat transfer at the nanoscale are still unclear. Recently, a study attempting to measure the local temperature profile using temperature-sensitive chemical reactions showed that the surface of isolated NPs may reach temperatures tens of degrees higher than the water bath.[14] However, a concern with the study was ignoring the effect of hydrodynamic radius during the heat transfer. Figure 3.2 shows with direct fluorescence

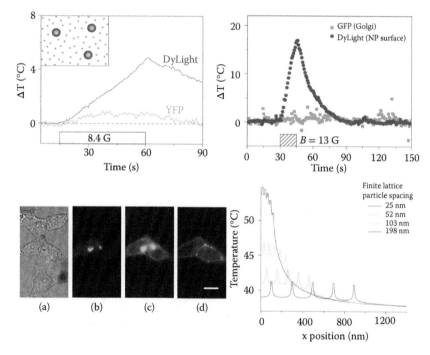

FIGURE 3.2

Temperature rise on the NP surface measured by fluorescence of DyLight549 conjugated to the streptavidin coating the NPs: on isolated NPs (top left) and on densely decorated cell membranes (top right). Bottom left: Specific targeting of the NPs on the cell membrane of HEK293T cells using surface protein presenting a biotin moiety which binds the streptavidin on the NPs. (a) DIC image, (b) GFP of Golgi marker, (c) CFP of surface marker, (d) red NPs. Bottom right: Computation of the local heating in the membrane for various NP spacings (water bath at 37°C.) (Unpublished and adapted from Huang, H. et al., *Nat Nanotechnol*, 5, 602–6, 2010.)

that the surface temperature of the NP may be a few degrees above the water temperature. However, in no case is there a detectable temperature rise at distances where a temperature-sensitive protein may be placed (~10 nm). We discuss the situation on the cell membrane in detail at the end of this chapter.

3.3 Experimental Considerations for Thermomagnetic Stimulation

3.3.1 NP Synthesis and Functionalization

As seen above, effective hyperthermia requires the NPs to possess high saturation magnetization. Yet, they should lack remnant magnetization (i.e., being superparamagnetic) to allow delivery of NPs to their *in vivo* targets without agglomeration. Among all ferrite materials suitable for bioapplications, manganese ferrite $MnFe_2O_4$ NPs have the highest magnetization, are stable against oxidation, and are often chosen as superparamagnetic core. Other common choices include magnetite Fe_3O_4, cobalt-iron-oxide $CoFe_2O_4$, and iron-platinum $FePt$.[15] Recently, a core–shell geometry consisting of a soft- and a hard-magnetic material was shown to provide higher magnetization and better heating as solid NP of the same size.[16] There are some commercial NP sources, but the majority of researchers synthesize their NPs themselves.

For optimal heating, the NPs should be monodisperse with high crystallinity. Figure 3.3 shows the theoretical specific absorption rate (SAR) as function of NP diameter for $CoFe_2O_4$-$MnFe_2O_4$ core–shell NPs for specific experimental parameters.[17] The optimal diameter for these particles would

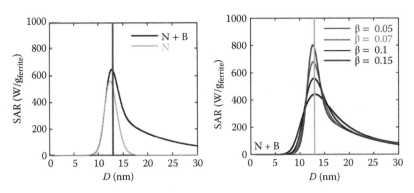

FIGURE 3.3

Simulations for specific absorption rate (SAR) as function of NP radius for $CoFe_2O_4$-$MnFe_2O_4$ core–shell NPs. (a) Comparison of Brownian (B) and Néel (N) relaxation contributions to the SAR as function of NP diameter D. (b) Effect of size dispersion on the SAR. ($T = 300°K$; $H = 22.5$ kA/m; $f = 425$ kHz; $M_s = 445$ kA/m.) (From Zhang, Q. et al., *Chem Mater.*, 27, 7380–7387, 2015.)

be 13 nm. If the size dispersion increases from 5% to 10%, the SAR drops by 35%. Such uniform NPs are best synthesized in organic solvents.[15,18] The resulting NPs are hydrophobic and require surface modification by either polymer ligand exchange or inorganic surface coating (silica or carbon).[19] Protocols for superparamagnetic NPs synthesis are found in other chapters in this book. The common criterion for optimal heating is that they should have very small size dispersion and the size should match the excitation frequency.

The NPs are made water dispersible by surface ligand exchange and coating with small hydrophobic molecule linked to a large hydrophilic molecule, either protein or polyethylene glycol (PEG). There are basically three methods: (1) enclosing in a lipid-PEG monolayer, (2) using a small linker such as 2,3-dimercaptosuccinic acid (DMSA) to attach coating proteins, or (3) encapsulating in a cross linked polymer with subsequent PEG modification. Alternatively, NPs have been enclosed in silica. All these coatings add substantially to the final hydrodynamic radius of the NPs, between 2 and 7 nm.

These functional coats are sometimes doped with a fluorophore for visualization of the NPs and local temperature sensing (see Section 3.3.2). Also, they provide the means to specifically target the NPs by covalently attaching a targeting protein. This may be an antibody against a surface expressed protein or a tag on that protein, e.g., his tag on the TRPV1his channel used to thermostimulation. Alternatively, it may use a click-chemistry approach or an avidin–biotin linkage. It is possible to genetically target cells of interest by expressing an engineered membrane marker protein AP-CFP-TM, containing a transmembrane domain (TM) and a biotin acceptor peptide (AP).[20,21] The biotin acceptor peptide is enzymatically biotinylated by the coexpressed BirA protein, forming specific binding sites for the streptavidin-conjugated NPs.

3.3.2 NP Physical and Magnetic Characterization

The final composition of the NPs should be analyzed inductively coupled plasma mass spectrometry (ICP-MS). Their size distribution and monodispersity of the NPs may be measured by transmission electron microscopy (TEM) before and after coating, and their hydrodynamic radius in aqueous suspension by dynamic light scattering (DLS).

To characterize the magnetic properties, the saturation magnetization should be measured at low temperature and at room temperature. Superparamagnetic NPs may be ferromagnetic and show a hysteresis at low temperature, but should have no hysteresis at room temperature. Performing this measurement continuously while heating the sample from low temperature to room temperature permits modeling the magnetic properties and fit anisotropy and magnetization. To be able to make predictions about the Néel and Brownian contributions to the heating, these measurements should be performed on two sample preparations: one dried and immobile sample and one sample in suspension. For ion-channel actuation, the NP will be immobilized on the cell and only the Néel relaxation will contribute to the heating.

3.3.3 NP Heating Capacity: SAR

The SAR quantifies the rate of energy deposition in tissue in hyperthermia, which is a measure of the amount of energy converted by the magnetic particles from the magnetic field into heat per unit time and mass. The SAR of a magnetic fluid is determined by the initial linear temperature rise of a fluid measured after switching on the magnetic field: $SAR = c \dfrac{dT}{dt}\Big|_{t=t_i}$, in which c is the heat capacity of the sample (Figure 3.4).[7] The original definition of the SAR is normalized to weight of the magnetic fluid as the power of the NP heating is used to heat up the surrounding fluid and the global fluid temperature is measured as function of time. Often, researchers will also report the SAR normalized the weight NP or to weigh magnetic material inside the NP in case the NPs contain significant amount of nonmagnetic material in their shell. All of these have advantages and disadvantages but it leads to some confusion. The SAR per fluid is preferred as it is directly related to the measurement but it does depend on the concentration of NPs. Figure 3.4 shows SAR measurements of several different types of NPs under the same condition. As the solution heats above room temperature, the measurement is not adiabatic anymore and the cooling has to be accounted for. Hence, the SAR is best calculated from the data within the first 0.5°K temperature rise. Figure 3.4 shows some curves from NPs whose SAR appears to increase with

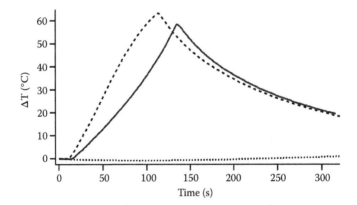

FIGURE 3.4
SAR of two core–shell NPs, $CoFe_2O_4$-$MnFe_2O_4$ (dashed line) and $MnFe_2O_4$-$CoFe_2O_4$ (solid line), and pure water (dotted line) measured ($T = 300°K$; $H = 22.5$ kA/m; $f = 425$ kHz; $M_s = 445$ kA/m). The $MnFe_2O_4$-$CoFe_2O_4$ appears to have an increasing SAR with increasing temperature because its blocking temperature is above room temperature. (From Huang, H. and A. Pralle, unpublished.)

temperature. This is the case for NPs whose magnetic blocking temperature lies about room temperature.

Theoretically, the rate of temperature increase in an aqueous dispersion is described by $\Delta T/\Delta t = P\ C/c_v$, where C is the volume concentration and c_v the volumetric specific heat. Therefor SAR is linearly proportional to the power loss per particle and the particle concentration: $SAR = P \cdot C / \rho$, in which ρ is the density of mass of the sample. Hence, the theoretically expected heating rate for the used NP dispersion at 0.01 vol% concentration would be 1 K/s, which is within the same order of magnitude as the measured value.

Small solenoid coils are used to apply the RF magnetic fields for the SAR measurements and cell stimulations. Two frequency ranges of magnetic fields are useful: 20–40 MHz is ideal for stimulating smaller NPs ($d \sim 6$ nm) while 400–500 kHz are optimal for large NPs (14–18 nm). The high-frequency regime has the advantage that the magnetic field and hence the current in the coil can be smaller, so that the coil does not require cooling. Hence, the coil can be very small, with small inductance and RF generators from the radio transmission field may use to build a simple set up. Magnetic field strength between 0.67 and 1 kA m^{-1} (8.4–13 G) may be generated with signal generator (Marconi Instruments), amplified by a 100-W amplifier (Amplifier Research). In the 400–500 kHz frequency range, there are commercial RF field sources because this is used for induction welding (MSI Automation, Wichita, KS; Ambrell, Rochester, NY; and others). As the frequency is 100-fold lower, the field has to be 10-fold higher, around 400 G. The currents in the coil are very large (10 seconds of Ampere) and require a water-cooled coil, which increases the volume of the coil and of the strain field complicating the integrating into a microscope.

3.3.4 Measuring Local Heating Using Ence Quantum Yield and Lifetime

In order to achieve stimulation through thermal activation of temperature-sensitive ion channels expressed on the plasma membrane of the cells, a method should be designed to raise the local temperature around the plasma membrane but leaving the inside of the cell minimally perturbed. It is therefore important to monitor local temperature changes at different compartments of the cells. While a conventional thermocouple can be used to measure heating in a bulk NP suspension, no readily available temperature probe can measure temperature changes within a cell, calling for a non-invasive, versatile, molecular-scale thermo sensor.

However, the fluorescence intensity and lifetime of chemical and biological fluorophores have a temperature dependence, which can be easily calibrated and utilized for molecular-scale temperature sensing.[7,22,23] Fluorescence is the emission of light by a molecule after it has absorbed a photon of a shorter wavelength. The absorbed energy excites the molecule to a higher energy state, from where it returns to the ground state either by emitting a photon or through nonradiative processes. Fluorescence quantum yield, Φ_F, defined as the ratio of the number of photons emitted to the number of photons absorbed, is a measure of the fluorescing efficiency of a specific fluorophore. It can also be defined as the ratio of radiative decay rate over the sum of all rates of excited state decay processes, $\Phi_F = k_F / \sum_i k_i$. One likely explanation for the decrease of the quantum yield Φ_F with increasing temperature is that with rising temperature, thermal vibrations within the molecule and environment intensify, increasing the rate of nonradiative processes.[7] The average time a molecule stays on the excited state before returning to ground state is called fluorescence lifetime. It is also the inverse of the total excited states decay rate, $\tau = 1 / \sum_i k_i$. Hence, the fluorescence lifetime also decreases with temperature increase.

Figure 3.5 shows that both the fluorescence intensity and lifetime of a fluorophore are temperature dependent, decreasing approximately linearly with increasing temperature within the temperature range of interest of this study (25°C–40°C, a larger temperature range would show exponential dependence [Arrhenius behavior]).[7] For all the fluorophores investigated in this study, which includes organic dyes, fluorescent proteins, and quantum dots, fluorescence intensity always decreases with increasing temperature, at a rate of around –1%/°K (Figure 3.5b).

With the temperature dependence calibrated, the fluorescence intensity of any fluorophore may be utilized for temperature sensing at molecular level inside life cells or organisms. After labeling specific parts of the cell with a calibrated fluorophore, one can obtain information about temperature changes at the location of the fluorophore simply by recording the fluorescence intensity over time. As fluorescent quantum yield and lifetime may also depend on other environmental parameters, such as pH and dielectric,

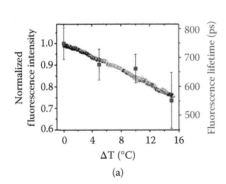

Fluorophore	$\Delta F(T)/F(T_0)$
DyLight549	–1.50 %/°K
Fluorescein	–1.15 %/°K
Rhodamine6G	–1.55 %/°K
Rhodamine6G in silica shell around mag. NP	–1.43 %/°K
ANNINE6 in membrane	–0.81 %/°K
Yellow fluorescent proteins (YFP)	–1.30 %/°K
Cerulean fluorescent proteins (C FP)	–1.65 %/°K

(a)   (b)

FIGURE 3.5
(a) Temperature dependence of fluorescence lifetime and of normalized fluorescence intensity of the small fluorophore DyLight549. (b) Table of the temperature dependence of some commonly used fluorophores and fluorescent proteins in various environments (water, membrane, or in a silica shell around manganese ferrite NPs). (Adapted from Huang, H. et al., *Nat Nanotechnol*, 5, 602–6, 2010.)

this has to be controlled in any application. Clearly, photo bleaching should be minimized and corrected before converting the intensity changes into temperature changes, or better the fluorescence lifetime should be used.

3.3.5 Local Temperature on NP Surface and in the Cell Membrane

As discussed above, the power dissipated from isolated NPs is likely too small to affect any protein when embedded in water as water is such a good heat conductor. Using fluorescence intensity from DyLight549 as the "thermometer," it is possible to record the temperature on the NP surface in dilute aqueous dispersions to determine the heating of isolated NPs and when arranged in a dense array on cells. The surface temperature of NPs was measured using the fluorescence for a dilute suspension of NPs (10 nM) (Figure 3.2, top left), and when attached to the cell surface (Figure 3.2, top right). In the suspension, the average NP–NP distance is 1 μm. Only negligible bulk heating was measured. The surface of the NPs reached 4°C over ambient, but that is still insufficient to affect a protein 10–15 nm away. Figure 3.2, top right, shows that regional heating of the cell membrane decorated with a high number of NPs can be quiet high. Within 15 seconds of applying the RF magnetic field, the cell surface temperature increased by more than 15°C. Because of the two-dimensional arrangement of the NPs, this temperature increase does not spread into the third dimension. The green fluorescent protein (GFP) fluorescence intensity in the Golgi remained virtually unchanged demonstrating that the heating is limited to the immediate plasma membrane vicinity (Figure 3.2, top right). Computations of the local heating for various NPs spacings on the cell membrane suggest that the NPs should be within 100 nm of each other to raise the average membrane temperature to 42°C if the water bath is at 37°C.

3.4 Applications

3.4.1 Remote Triggering of Cellular Calcium Influx

Figure 3.1 summarizes the principle of thermomagnetic cell stimulation, which couples the local NP heating to the opening of TRPV1 ion channel causing calcium influx into the cell. This calcium signal then triggers further events depending on the cell type and genetic engineering performed. In order to effectively heat the TPV1 channels to stimulate specific cells *in vivo*, a high local density of NPs is needed to cause significant regional heating, i.e., along the membrane surface. Cell specificity and high NP loading may be achieved by genetically manipulating the cells to express a biotinylated membrane protein to which the streptavidin-conjugated NPs bind. Alternatively the NPs may be coated with an antibody and directly targeted to the temperature-sensitive ion-channel TRPV1.[11] The former provides the advantage of separate control over the concentration of NP binding sites and ion channels; the later tethers the NP potentially closer to the channel.

HEK293 cells specifically labeled with streptavidin-coated NPs, and expressing the TRPV1 channel and a calcium sensor, show rapid (within seconds) calcium influx upon application of the RF magnetic field.[7] The increase is caused by calcium influx through thermally activated TRPV1 channels as cells with NPs but without TRPV1 channels, as well as cells with TRPV1 channels but without NPs, did not show any calcium influx upon application of the same magnetic field. The calcium influx is comparable to the influx caused by opening the channels with the agonist capsaicin.

3.4.2 Remote Triggering Action Potentials and Behavior

The magnetic field-induced calcium influx results in neuronal depolarization sufficient to elicit an action potential, which is necessary for the control of neuronal function. Figure 3.6 shows the change of the membrane potential of cultured hippocampal neurons expressing TRPV1 and labeled with NPs. After applying the RF magnetic field, the fluorescence intensity

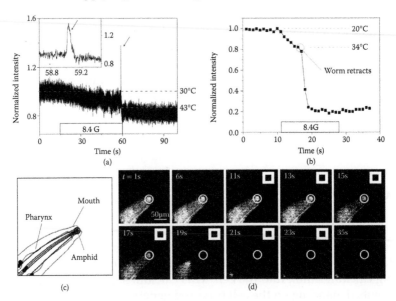

FIGURE 3.6
Action potentials elicited in TRPV1-expressing hippocampal neurons labeled with NPs and subject to alternating magnetic field. Membrane potential was recorded using voltage-sensitive dye, ANNINE6. Remote stimulation of thermal avoidance response in *C. elegans*. NPs coating the amphid region (a, c) of the nematode raises the temperature above thermal avoidance threshold (b) and triggers a retraction (d, 19 seconds).

of the membrane voltage ANNINE6 decreases as the membrane temperature rises. When the membrane temperature had reached about 40°C, the voltage-sensitive dye registered several small membrane voltage spikes followed by an action-potential type depolarization (Figure 3.6a insert).

The evoked action potentials are strong enough to alter the behavior of a model organism. *Caenorhabditis elegans* is often used as model organism because it contains only 303 neurons whose connective wire diagram is known. Figure 3.6b and d show how the thermomagnetic stimulation evokes an avoidance reaction.[7] PEG-phospholipid coated NPs fed to the nematodes enriched in the mucus layer protecting the amphid region where the sensory neurons end. The image sequence at the bottom of Figure 3.6 shows the response of a being partially anesthetized by being mounted on a thin agarose pad containing 9-mM NaN_3. After application of the field (indicated by the square in the picture), the NP fluorescence decreased, indicating heating, until at a temperature of 34°C the worm retracted. This threshold agrees with a previously found noxious thermal avoidance at 33°C.[24] Freely crawling *C. elegans* showed the same avoidance only when NPs and magnetic field were present. The similar response of all the worms within the magnetic field demonstrates a major advantage of this method: it does not require focusing the stimulating energy field onto one cell or animal, which could be challenging for stimulating a large group of animals or cells within one organism simultaneously.

3.4.3 Remote Activation of Protein Production and Secretion

The thermomagnetic stimulation of cellular calcium influx may be used to control other cell functions as well. Calcium is a common messenger in cell signaling whose precise signal depends on the extent and duration of the calcium increase. Thermomagnetic TRPV1 channel activation provides a long tonic elevation of calcium levels which directly leads to membrane depolarization as shown above. It may also control slow processes via calcium binding to regulatory proteins. Sara Stanley and coworkers engineered calcium dependence into gene expression of insulin.[11] They were able to used NP heating in an RF magnetic field to remotely activate insulin production *in vivo*.[11] The NPs were targeted either directly to modified TRPV1 channels or to biotin-anchor proteins as above. The same cells were engineered to contain a gene for insulin under a series of calcium-dependent promotors (Figure 3.7). The induced calcium influx stimulates synthesis and release of the bioengineered insulin. This process required 30 minutes of RF heating. By engineering the stimulation and insulin production into tumor cells and implanting this tumor in mice, the researchers were able to show that the produced insulin was functional, lowering the blood glucose levels.

Remote control of hormone or protein production is a good application of the technique at its current stage, in the sense that insulin release is a

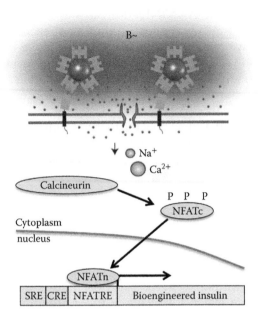

FIGURE 3.7
Schema of NP-induced cell excitation and insulin expression. The RF-induced local NP heating opens temperature-sensitive TRPV1 channels. Calcium entry activates calcineurin, leading to dephosphorylation of NFAT and translocation to the nucleus. NFAT binding initiates
gene expression of a bioengineered human insulin gene. Additional calcium-dependent signal
pathways stimulate gene expression via binding to SRE and CRE. (Adapted from Stanley, S.A.
et al., *Science*, 336, 604–608, 2012.)

relatively slow process that does not require high temporal precision. One
can foresee that upon improvement of the heating efficiency and stimulation speed, the technique can be used in controlling other calcium regulated
processes, such as muscle contractions.

3.5 Summary and Conclusions

The reviewed remote control of ion channels in cells using RF magnetic
field heating of NPs provide the great advantage over other approaches
to be completely contactless and to work deep inside the body. It may be
applied broadly to calcium controlled signaling and may be extended to
others by engineering other thermally sensitive proteins. First applications
have shown action potential and behavior triggering and the control of hormone (insulin) production *in vivo*. At present, the method achieves response
times of tens of seconds, which is useful for the control of protein production
and for tonic nerve stimulation. However, for precise neurostimulation, the

response time needs to be improved, which may be achieved by optimizing the NPs, their targeting to the cell, and maximizing the magnetic field. Each of these optimizations may yield enough improvement to achieve a subsecond response time. To improve the stimulation speed, the heating power of NPs needs to be optimized, which can be achieved by optimizing NP composition and anisotropy, improving monodispersion by better control of size distribution during synthesis, and optimizing the pairing between NP size and RF magnetic field frequency.[12]

NP delivery, stability, and cellular uptake *in vivo* are still challenges to this stimulation technique and are areas of active research worldwide. Stanley and coworkers showed that repeated experiments over several days were only possible when more NPs were injected.[11] An attractive alternative to external NP delivery may be the genetic encoding of NP synthesis in the target cells. The simplest NP could be the iron-storage protein ferritin. Some groups have reported that RF application to ferritin expressing cells *in vivo* has produced the desired effect, insulin release,[11] or apoptosis.[25] However, most groups, including ours, have failed to measure any appreciable heating from human or horse ferritin. This is the fact that ferritin was not localized near the ion channels, shed doubt on the *in vivo* results being thermostimulation. Iron or cobalt leaking from the ferritin cage would be toxic and make the cell leaky, which may explain the observations. Still, creating a genetically encoded NP capable of thermostimulation would remove a major technical hurdle to *in vivo* application of this method. The best heating genetically encoded NPs are currently magnetosomes extracted from magnetotactic bacteria.[26] There is some promise in eventually transferring the genetic pathway for magnetosome synthesis.[27]

References

1. Banghart, M. et al., Light-activated ion channels for remote control of neuronal firing. *Nat Neurosci*, 2004. **7**(12): pp. 1381–6.
2. Zemelman, B.V. et al., Selective photostimulation of genetically chARGed neurons. *Neuron*, 2002. **33**(1): pp. 15–22.
3. Boyden, E.S. et al., Millisecond-timescale, genetically targeted optical control of neural activity. *Nat Neurosci*, 2005. **8**(9): pp. 1263–8.
4. Fenno, L. et al., The development and application of optogenetics. *Annu Rev Neurosci*, 2011. **34**: pp. 389–412.
5. Hughes, S. et al., Selective activation of mechanosensitive ion channels using magnetic particles. *J R Soc Interface*, 2008. **5**: pp. 855–63.
6. Mannix, R.J. et al., Nanomagnetic actuation of receptor-mediated signal transduction. *Nat Nanotechnol*, 2008. **3**(1): pp. 36–40.
7. Huang, H. et al., Remote control of ion channels and neurons through magnetic-field heating of nanoparticles. *Nat Nanotechnol*, 2010. **5**(8): pp. 602–6.

8. Caterina, M.J. et al., The capsaicin receptor: A heat-activated ion channel in the pain pathway. *Nature*, 1997. **389**(6653): pp. 816–24.

9. Tominaga, M. et al., The cloned capsaicin receptor integrates multiple pain-producing stimuli. *Neuron*, 1998. **21**(3): pp. 531–43.

10. Lima, S.Q. and G. Miesenbock, Remote control of behavior through genetically targeted photostimulation of neurons. *Cell*, 2005. **121**(1): pp. 141–52.

11. Stanley, S.A. et al., Radio-wave heating of iron oxide nanoparticles can regulate plasma glucose in mice. *Science*, 2012. **336**(6081): pp. 604–8.

12. Rosenzweig, R.E., Heating magnetic fluid with alternating magnetic field. *J Magn Magn Mater*, 2002. **252**: pp. 370–4.

13. de Chatel, P.F. et al., Magnetic particle hyperthermia: Neel relaxation in magnetic nanoparticles under circularly polarized field. *J Phys Condens Matter*, 2009. **21**(12): p. 124202.

14. Riedinger, A. et al., Subnanometer local temperature probing and remotely controlled drug release based on azo-functionalized iron oxide nanoparticles. *Nano Lett*, 2013. **13**(6): pp. 2399–406.

15. Sun, S. et al., Monodisperse FePt nanoparticles and ferromagnetic FePt nanocrystal superlattices. *Science*, 2000. **287**(5460): pp. 1989–92.

16. Lee, J.H. et al., Exchange-coupled magnetic nanoparticles for efficient heat induction. *Nat Nanotechnol*, 2011. **6**(7): pp. 418–22.

17. Zhang, Q. et al., Model driven optimization of magnetic anisotropy of exchange-coupled core–shell ferrite nanoparticles for maximal hysteretic loss. *Chem Mater*, 2015. **27**(21): pp. 7380–7.

18. Park, J. et al., Ultra-large-scale syntheses of monodisperse nanocrystals. *Nat Mater*, 2004. **3**(12): pp. 891–5.

19. Erathodiyil, N. and J.Y. Ying, Functionalization of inorganic nanoparticles for bioimaging applications. *Acc Chem Res*, 2011. **44**(10): pp. 925–35.

20. Howarth, M. et al., Targeting quantum dots to surface proteins in living cells with biotin ligase. *Proc Natl Acad Sci U S A*, 2005. **102**(21): pp. 7583–8.

21. Howarth, M. and A.Y. Ting, Imaging proteins in live mammalian cells with biotin ligase and monovalent streptavidin. *Nat Protoc*, 2008. **3**(3): pp. 534–45.

22. Karstens, T. and K. Kobs, Rhodamine B and rhodamine 101 as reference substances for fluorescence quantum yield measurements. *J Phys Chem*, 1980. **84**(14): pp. 1871–2.

23. Duhr, S. et al., Thermophoresis of DNA determined by microfluidic fluorescence. *Eur Phys J E Soft Matter*, 2004. **15**(3): pp. 277–86.

24. Wittenburg, N. and R. Baumeister, Thermal avoidance in *Caenorhabditis elegans*: An approach to the study of nociception. *Proc Natl Acad Sci U S A*, 1999. **96**(18): pp. 10477–82.

25. Fantechi, E. et al., A smart platform for hyperthermia application in cancer treatment: Cobalt-doped ferrite nanoparticles mineralized in human ferritin cages. *ACS Nano*, 2014. **8**: pp. 4705–19.

26. Hergt, R. et al., Magnetic properties of bacterial magnetosomes as potential diagnostic and therapeutic tools. *J Magn Magn Mater*, 2005. **293**(1): pp. 80–86.

27. Kolinko, I. et al., Biosynthesis of magnetic nanostructures in a foreign organism by transfer of bacterial magnetosome gene clusters. *Nat Nanotechnol*, 2014. **9**(3): pp. 193–7.

4

Magnetic Twisting Cytometry

Daniel Isabey, Christelle Angely, Adam Caluch,
Bruno Louis, and Gabriel Pelle

CONTENTS

4.1 Introduction 59
4.2 Historical Aspects 61
4.3 Methodological Aspects 63
    4.3.1 Experimental Device 63
    4.3.2 The Bead as Cell Twister and Transducer 66
    4.3.3 Microrheological Viscoelastic Models 67
    4.3.4 Assessment of Mechanical and Adhesion Properties
          from Transient MTC Loading 71
    4.3.5 Molecular Adhesion Models 72
    4.3.6 MTC Application to Characterization of a Living Cell
          Response 79
    4.3.7 Some Remarkable Features of MTC 83
4.4 Conclusions and Perspectives 85
References 86

4.1 Introduction

Living cells in the human body are constantly subjected to mechanical stimulation throughout life. The origin of stresses and strains can arise from both the external environment and internal physiological conditions. Thereby, cells can respond to mechanical stimulation in a variety of ways depending on the magnitude, direction, and distribution of mechanical stimuli. To understand how cells mechanically respond to physical loads, there is a need to control the mechanical stress exerted on studied cells. This can be achieved ex vivo, in living cell experiments. A number of micromanipulation techniques have been purposely developed, notably magnetic twisting cytometry (MTC), micropipette aspiration, optical tweezers (OT), atomic force microscopy (AFM) indentation, cytoindentation or microplaques, and fluid shear flow. MTC, invented two decades ago by Wang et al. (1993), considers that the cell's mechanical response is dictated by its intracellular architecture, which is intentionally probed by twisting ferromagnetic microbeads bound to transmembrane receptors physically linked to this intracellular structure. Eukaryotic cells contain such an intricate molecular framework, the cytoskeleton (CSK), composed of interconnected microfilaments, microtubules, and intermediate filaments that respond but also generate mechanical loads, and thereby are largely responsible for the cell's ability to resist shape distortion. Cellular integrity and mechanical properties lie in the balance between tensional and compressive forces (Chicurel et al., 1998). The latter are constantly redistributed throughout the whole architecture, that is, the proper of "tensegrity" structures (Ingber, 1993; Stamenovic and Coughlin, 1999; Wendling et al., 1999) and are capable of efficiently transmitting mechanical stress and activating cytoplasmic (and transmembrane) proteins throughout the cell (Hu et al., 2003; Na et al., 2008). The corollary, used by MTC, is that probing directly the CSK enables us to locally capture the essentials of the cell's mechanical properties while minimally affecting the measured properties (Féréol et al., 2008). However, cellular viscoelastic properties obtained with beads having the same coating and size but different techniques (e.g., OT and MTC) are not so easy to reunify (Laurent et al., 2002b). Magnetic fields are advantageously used because they are robust, easy to control, and can respond rapidly while a huge range of frequencies can be tested (e.g., from 10^{-3} to 10^4 Hz; Chowdhury et al., 2008). The MTC method has been modified to be coupled with microscopy visualization including high-resolution microscopic imaging methods. This "high-frequency MTC" enables us to characterize the dynamic cell mechanical behavior over large populations of beads and cells (Fabry et al., 2001). In such a case, an averaged nanoscale axial displacement is measured instead of an averaged bead rotation measured in the initial method (Wang et al., 1993). Consistent with the soft glass rheology (SGR) model, the dependence of cell stiffness on loading frequency has been represented by a unique weak power law (Fabry et al., 2001). However, a dynamic rheology model which takes into account multiple parallel noncovalent bonds reveals that two power laws are necessary, which is indeed the case when cells are tested over a wider range of loading frequencies (Chowdhury et al., 2008). Thus, a comprehensive cellular model capable of relating the cell mechanical response to the complex cellular and molecular structures still remains to be elaborated. This lack of comprehensive cellular models can be associated with the constant difficulty of relating the cell mechanical response experimentally measured to the underlying cellular and molecular structures (Stamenovic et al., 2007).

Variability in mechanical properties reflects not only the differences among experimental conditions, which are specific to each technique, but more fundamentally the spatial complexity and multiscale time dependence of cellular and molecular structures (Fabry et al., 2001; Hu et al., 2003; Ingber, 1998). Indeed, it is now widely accepted that structural and molecular aspects dominate when looking at living cell mechanical responses. The role of molecular motors, such as actomyosin, has been largely suspected in stretched cells (Féréol et al., 2008) and especially in contractile cells, that is, muscle cells (Mitrossilis et al., 2009), but it is also at the origin of intracellular tension that is more or less present in all cell types (Wang et al., 2002; Wendling et al., 2000). However, the contribution of adhesion molecules was mostly ignored when looking at the cell mechanical response. We are presently showing that an appropriate analysis of data obtained by MTC enables the revelation of, in addition to CSK mechanical properties, some hidden aspects of the kinetics of cell–matrix adhesion. Indeed, the partial defect in integrin–matrix attachment extracted from the time-dependent signal of bead rotation constitutes a unique opportunity to reveal the molecular properties at the cell–matrix interface.

It has also been demonstrated that mechanical forces can alter the chemical activities of a number of CSK-related molecules (Chicurel et al., 1998; Meyer et al., 2000; Wang et al., 1993; Yoshida et al., 1996). This is susceptible to modifying the biochemical reactions that mediate critical cellular functions. Little is known about the mechanisms by which individual cells sense mechanical signals and transduce them into changes in intracellular biochemistry and gene expression—a process that is known as mechanotransduction. Incidentally, twisting beads have been used in a number of studies on mechanical force-induced signaling molecules during recent years (Hu et al., 2003; Na et al., 2008). Using data obtained in living cells, we show in this chapter and in a recent publication (Isabey et al., 2016) how the MTC method can reveal the effect of stress or strain at very different scales.

4.2 Historical Aspects

Applying mechanical forces and stresses to cells by means of magnetic beads is surprisingly not a new idea as it dates from the beginning of the twentieth century. Magnetic dragging and twisting were first described from the fundamental point of view of physics by Francis H. C. Crick, an English molecular biologist, physicist, and neuroscientist who received the Nobel prize in 1962 for his pioneering work on 3D structure of DNA. He published in the 1950s, with A. F. W. Hughes, a couple of reference papers showing that external magnetic fields can be used to twist magnetic particles within isolated cells and probe the viscous properties of the cytoplasm (Crick and Hughes, 1950). More than two decades later, David Cohen from MIT (the inventor of magneto-encephalography) was the first to detect the presence of inhaled magnetic material within the human body while he was monitoring physiological magnetic fields from cardiac electric currents. He demonstrated that contaminating particles retained in the lungs (and other organs) due to dust inhalation could be measured after a permanent magnetic moment was induced by brief application of a strong magnetic field (Cohen, 1973). The combined effect of these aligned moments was to produce a remanent field that was detected outside the patient with a sensitive magnetometer. This remanent field has been initially proposed to quantify the amount of contaminating particles in the body, such as finding the magnetic material deposited in workers occupationally exposed to airborne magnetic dusts (Freedman et al., 1980). This technique has been called "magnetopneumonography." The most striking observation made by Cohen was that the remanent field from the lungs decreased by as much as a factor of six during the first hour after magnetization, a phenomenon he called "relaxation" (Cohen, 1973), which has been latter confirmed by many studies such as those of Peter Valberg, a physicist from the Harvard School of Public Health (Valberg and Brain, 1979). Since the particles' magnetic moments are permanent while the remanent field can be restored to its initial value by remagnetization, Cohen considered that this "relaxation" was due to some process capable of randomly rotating the magnetic particles retained in the lungs. To target the cellular mechanisms responsible for particle motion among a number of potential little-controlled in vivo mechanisms, experiments were performed in cultured macrophages obtained by lung lavages in animals who initially inhaled magnetic particles (Valberg, 1984; Valberg and Albertini, 1985). Based on these ex vivo cellular studies as well as on in vivo studies in animals in which the decay in magnetic signal could be reproduced (Brain et al., 1984), the "relaxation" phenomenon could be attributed to macrophage cytoplasmic motions and thus related to phagocytic activity.

The idea of using intracellular magnetic particles as nonoptical probes of motility and rheological properties of the cytoplasm (mainly of pulmonary macrophages) was initiated by two physicists of the Harvard School of Public Health: Valberg and Butler (1987). They described the physical principles of a new technique called "cell magnetometry," and developed an appropriate hydrodynamic model to deduce the cytoplasmic viscosity from the resistance to particle motion. These authors defined many aspects of the techniques that are today still in use in the MTC system in our laboratory: (1) magnetic particle rotation was measured with a sensitive fluxgate magnetometer and (2) the particles' remanent magnetic field of the order of 1 nT ($=10^{-9}$ T) was measured in presence of a large twisting field of about 5 mT (1 mT = 10^{-6} T). The latter condition required minimizing external magnetic noise with mu-metal shielding and rotating the macrophage culture in order to use lock-in amplification of the rather small remanent magnetic signals. Interestingly, the link of cell magnetometry with the intracellular structure was not clear except that incubation of macrophages with cytochalasin D tended to inhibit particle intracellular motion in a dose-dependent manner (Valberg and Feldman, 1987). Moreover, cell magnetometry was presented as a useful tool to characterize intracytoplasmic cell properties (i.e., essentially viscosity of alveolar macrophages) not only for in vitro experimental conditions but also for in vivo studies.

Measuring mechanical properties in a variety of cells that do not phago-cytose the magnetic beads is a biologically relevant idea of Ning Wang and Donald Ingber who proposed attaching spherical ferromagnetic particles to the CSK-specific cell surface receptors normally expressed on the apical face of adherent cells (Wang et al., 1993). This new probing system associated with the relative easiness of manipulating magnetic fields on cell culture has constituted an improvement which has enabled MTC to become a leading technique for the study of cellular responses to mechanical cues in many biological areas. Almost all cell types can now be measured using magnetic beads. We presently demonstrate that the strength of the physical linkage between ligand-receptors-CSK can now be probed by MTC. Accordingly, liv-ing cell results can be reanalyzed by such appropriate adhesion models to characterize the bead–cell interface by molecular dynamics. Altogether, we presently bring new arguments that confirm the richness of the bead twist-ing method to characterize not only the multiscale CSK structure but also cell–matrix interactions.

4.3 Methodological Aspects

4.3.1 Experimental Device

By principle, MTC uses ferromagnetic microbeads to apply twisting forces, that is, essentially shear stresses, to specific transmembrane receptors ini-tially present at the surface of living cells (Wang et al., 1993). MTC has been extensively described in previous reviews or synthesis articles on cell mechanics (Féréol et al., 2008; Lele et al., 2007). We present hereafter a summary.

Carboxylated ferromagnetic beads of about 4.5 μm in diameter (Spherotech) are coated with a ligand (e.g., synthetic RGD peptide from the cell-binding region of fibronectin, receptor-specific antibodies, ECM molecules) (Figure 4.1a). Before experiments, the coated ferromagnetic beads are added to the cells (~2 beads/cell) in a serum-free, chemically defined medium, and incubated for 10 minutes at 37°C, followed by gentle washing three times with PBS to remove unbound beads prior to force application.

MTC experiments are performed over large populations of living adher-ent cells cultured in wells of about 6 mm in diameter (Figure 4.1a). To apply twisting forces to beads and bound surface receptors, a strong (~150 mT) and brief (~150 ms) magnetic field is applied to the receptor bound ferro-magnetic beads using horizontal Helmholtz coils (Ohayon et al., 2004). This aligns the magnetic dipoles of the beads in the same horizontal direction. Then, a weaker transient (plateau duration ~60 seconds in between two ramps of duration ~400 ms) magnetic field $\bar{H}$ (≤6 mT equivalent to 4% of

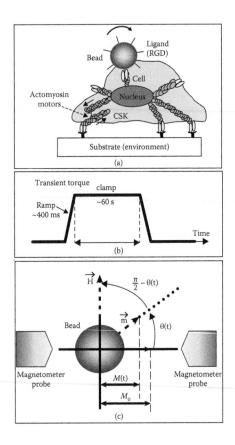

FIGURE 4.1

Magnetic twisting cytometry. (a) Living cells are cultured in treated small plastic wells (6-mm diameter) and incubated with bound ferromagnetic beads for a sufficient time (≥30 minutes) allowing the cells to engulf the beads. Beads are coated with the synthetic RGD peptide from the cell-binding region of fibronectin, anti-integrins antibody, or ECM molecules. These coatings are recognized by transmembrane integrin receptors that are also physically coupled to the intracellular CSK structure and primarily the actin filament network. After magnetization of beads by a strong magnetic field created by Helmholtz coils, magnetic dipoles are aligned in the horizontal direction and a weaker magnetic field (≤6 mT) also created by Helmholtz coils is applied in the perpendicular direction during 60 seconds. The resulting magnetic torque is responsible for bead twisting. The twisting field, adjustable in magnitude by changing the current intensity in Helmholtz coils, allows application of shear forces directly to the bound receptors and beyond to the CSK. (b) A transient trapezoidal torque signal is generated. It is constituted by a positive 400-ms ramp, a 60-second clamp with almost constant torque and terminated by a negative 400-ms ramp. (c) Before, during, and after bead twisting, the remanent magnetic field is measured by an on-line magnetometer whose probes are located in the horizontal plane (culture plane). $M(t)$ is the instantaneous magnetic field, M_0 is the remanent magnetic field measured before torque application. The bead deviation $\theta(t)$ is the angle between the horizontal direction and the bead magnetic moment $\vec{m}$. The complementary angle $\left[\dfrac{\pi}{2} - \theta(t)\right]$ is the angle between the bead magnetic moment $\vec{m}$ and the vertical direction of the perpendicular twisting field $\vec{H}$. The machanical and adhesion properties are computed based on knowledge of twisting field and angular strain calculated from the arc cosine of the $M(t)/M_0$.

the magnetization field) (Figure 4.1b) is applied in the perpendicular direction using vertically oriented Helmholtz coils (Figure 4.1c). As a result, the beads are twisted as the magnetic vector of the bead attempts to align to this "twisting field," thus applying shear force directly to the bound receptors. The resulting magnetic torques are in the range 450–1,200 pN × μm (Laurent et al., 2002b).

$$\vec{C} = \mu_0\,\vec{m} \wedge \vec{H}, \text{ i.e., } C = \mu_0\,m\,H\,\sin\!\left(\frac{\pi}{2}-\theta_{eq}\right) (\text{with } m \approx 2.3\times10^{-13}\,\text{Am}^2) \quad (4.1)$$

Here, m is the modulus of the bead magnetic moment obtained after calibration and μ_0 the permeability of free space. The spatially uniform perpendicular field H can be modified by changing the current in the Helmholtz coils, that is, up to a maximum $I \approx 15$ A. θ_{eq} is the bead deviation angle at the equilibrium between shear stress and CSK elastic recoil, that is, after 60 seconds of torque application.

Note that MTC measurements cannot be performed when the projected values of the remanent magnetic field are too small, for example, for values below the noise level of the magnetometer and probes (Féréol et al., 2008). A density of 2–10 beads per cell, that is, 20,000–50,000 beads on a cell culture, results in a maximum remanent field B_0 in the range 1–2 nT, which corresponds to a reasonable signal to noise ratio. This is why applied torque cannot be less than 400 pN × μm to obtain a measurable change in remanent field. On the other hand, to avoid excessive heating in the Helmholtz coils, torque cannot exceed 1,200 pN × μm. A constraint of the MTC system is to measure remanent magnetic fields in the horizontal direction while twisting magnetic fields 10^8 greater are superimposed (Figure 4.1c). This has been solved in prior bead twisting studies (Valberg and Butler, 1987) by rotating the well (Petri dish) containing beads and cells at, for example, 10 Hz, and using lock in amplification in addition to μ-metal shielding to screen external magnetic fields.

The resulting CSK deformation or angular strain induced by the twisting field is measured based on the mean bead rotation angle which is continuously measured (throughout 60 seconds of twisting period followed by 60 seconds of recovery period) from the change in projected magnetic bead moment, using the arc cosine of the ratio of remanent field to the field at time 0 (Figure 4.1c):

$$\theta(t) = \text{arc}\,\cos\!\left(\frac{M(t)}{M_0}\right) \quad (4.2)$$

This procedure enables measurement of an average time-dependent angular strain in response to the magnetic torque applied. The MTC method covers an intermediate range of maximum cellular deformation (bead rotation angles: 15°–60°) and, due to the large number of probing beads, provides homogenization of the cell response which is of particular interest to characterize a material whose behavior happens to be nonlinear when stress or strain increases (Ohayon et al., 2004).

Because the applied magnetic torque used in standard MTC measurements is roughly a trapezoidal signal (Figures 4.1b), the cellular response resembles a creep function which suggests that the CSK behaves as a viscoelastic solid-like system (see paragraph 3 and Figure 4.2). Before describing how to assess the biomechanical parameters characterizing the CSK viscoelastic behavior, it is necessary to consider the effective stress, which for geometrical reasons does not correspond to the apparent stress in the case of partial bead immersion.

4.3.2 The Bead as Cell Twister and Transducer

The cell responds to the bead by a variable degree of endocytosis which depends on cell phenotype. For a given torque, bead displacement in a cell

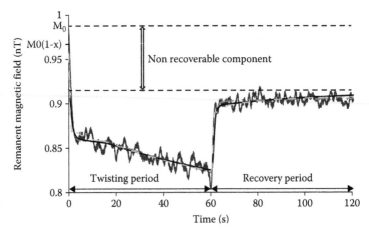

FIGURE 4.2

Signal of the remanent magnetic field in MTC. The instantaneous magnetic field $M(t)$ (in nT) is plotted versus time (in second) in living cells. This recording shows the response of Human Pulmonary Microvascular Endothelial Cells (HPMECs) to a 60-second twisting period followed by a 60-second recovery period. From an initial value M_0 of the order of 1 nT, we can detect an immediate (time independent) decay up to the value $M_0(1-x)$ in which x represents the proportion of free rotating beads. It is followed by a time-dependent decay in $M(t)$ used to analyze de-adhesion phenomena, which contributes in addition to the time-independent decay to the nonrecoverable component of the magnetic signal. The recoverable component of the magnetic signal is basically used to analyze the "global" viscoelastic response of the CSK from the different models (single-Voigt model, double-Voigt model, time-invariant power law) whose parameters, obtained by best curve fitting, lead to three creep functions presently superimposed to the experimental signal for comparison. Differences between the experimental curve and the three models are rather small.

is facilitated when the bead is only partially immersed (Féréol et al., 2008). Calling α the half-angle of the cone limiting the bead contact area and estimating α from 3D reconstruction of cortical actin images, $\alpha \approx \pi/3$ was measured in alveolar epithelial cell lines (A549) (Ohayon et al., 2004), $\alpha \approx 2\pi/3$ in alveolar macrophages (Féréol et al., 2006), $\alpha \approx \pi$ in human pulmonary microvascular endothelial cells (HPMVECs) (Wang et al., 2012). The electron microscopic image of a similar bead attached to a smooth muscle cell (SMC) (Fabry et al., 2003) reveals indeed much smaller angles of bead immersion: $\alpha \leq \pi/12$.

Static elasticity theory in continuous medium can be used to determine the role of α in the determination of the measured Young's modulus (E) of a cellular medium (Laurent et al., 2002b). When the bead is in equilibrium between the magnetic torque (Equation 4.1) and a reacting elastic torque C_e, the following relationships can be obtained: $C_e = -(8/3)\pi R^3 \, E \, \theta_{eq}$ for full immersion of the bead in infinite medium ($\alpha = \pi$), $C_e = -(4/3)\pi R^3 \, E \, \theta_{eq}$ for half immersion of bead, ($\alpha = \pi/2$), $C_e = -(4/3)\pi R^3 \, (\sin^3\theta) E \, \theta_{eq}$ for a small circular contact area $\alpha \leq \pi/6$. These analytical expressions explain for instance why as the same stress is applied through different-sized magnetic beads (e.g., 4.5-µm vs. 1.4-µm diameter), larger beads appeared stiffer or more resistant to rotation than smaller beads, a property that could be verified experimentally (Wang and Ingber, 1994). Half immersion corresponds to the standard definition of apparent elastic modulus used in several early studies (Planus et al., 1999; Pourati et al., 1998; Wang, 1998), that is, $E_{app} = \dfrac{C_e}{V_{bead}\theta_{eq}}$, in

which $\dfrac{C_e}{V_{bead}} (= \sigma_{app.})$ represents the resulting apparent stress exerted on the cell. Using this definition and a bead volume of ~50 µm³, the resulting stress exerted to the cell through the bead would vary between 8 and 24 Pa for the range of magnetic torques indicated above. Using this apparent value, E_{app} as reference value for cellular stress, the correcting factor: $\kappa = \dfrac{E_{app}}{E} = \dfrac{\sigma_{app}}{\sigma}$, can be calculated by numerical model (Ohayon et al., 2004). κ ranges from values below than 0.1 at low bead immersion to values approaching 2 for complete bead immersion. The discrepancy between apparent and corrected elastic properties is therefore maximal for small bead immersions. The above apparent stresses thereby result in significantly higher effective cellular stresses which stay in the range 8–200 Pa. In addition to the bead immersion angle α, cellular height and substrate properties which act as boundary conditions may also influence the torque-bead displacement relationships (Ohayon and Tracqui, 2005), for example, substrate properties might influence the 4.5 µm-microbead deviation for cell height below 7–8 µm (Féréol et al., 2006).

4.3.3 Microrheological Viscoelastic Models

The microrheological response of living cells locally twisted by ferromagnetic microbeads, that is, transiently exposed for 60 seconds to an almost constant stress (σ) magnetic torque and then relaxed, can be described in the course of time by a relationship between strain and stress, which takes into account that angular deformation, $\theta(t)$, is caused by the total history of the loading, up to the time t (Figure 4.2):

$$\theta(t) = f(t)\sigma(0) + \int_0^t f(t - t')\frac{d\sigma}{dt}dt' \tag{4.3}$$

$f(t)$ is the creep function (i.e., the strain generated by a step stress, normalized by a constant stress value).

Three types of microrheological models of cellular response were tested:

1. The viscoelastic solid made of a single-Voigt model:

$$f(t) = \frac{1}{E} + \frac{t}{\eta} \tag{4.4}$$

E and η are the elastic modulus (or cellular stiffness) and viscosity (or frictional) modulus, respectively, characterizing the response of a viscoelastic solid with a unique relaxation time constant: $\tau = \dfrac{\eta}{E}$

2. The viscoelastic solid made of a double-Voigt model:

$$f(t) = \frac{1}{E_1} + \frac{t}{\eta_1} + \frac{1}{E_2} + \frac{t}{\eta_2} \tag{4.5}$$

E_1, η_1 or E_2, η_2 are the elastic modulus (or cellular stiffness) and the viscosity (or frictional) modulus of the two components characterizing the viscoelastic solid response of a series of rapid and slow elements with, respectively, fast and slow relaxation time constants: $\tau_1 = \dfrac{\eta_1}{E_1}$ and $\tau_2 = \dfrac{\eta_2}{E_2}$ with, for example, $\tau_1 < \tau_2$.

The two elementary viscoelastic models above consider cell elastic behavior to be linear within a given MTC experiment, that is, for a given loading level. However, MTC experiments performed at different loading levels may reveal a stress-dependent behavior of cellular stiffness which was explained by different mechanisms at the cellular scale, for example, structural effects of the tensegrity type (Wendling et al., 1999), hyperelasticity of the cell material (Ohayon et al., 2004), or other mechanisms at the nanometer scale in relation with molecular actomyosin motors (Cai et al., 1998; Mizuno et al., 2007).

3. The time-invariant power law:

It is assumed here that $f(t)$ behaves as a power law of time:

$$f(t) = A_0 \left(\frac{t}{t_0}\right)^\alpha \tag{4.6}$$

Here, t_0 is an arbitrary reference time chosen for convenience equal to 1 second. From the parameters of the power law, that is, the exponent α and the prefactor A_0, it is also possible to calculate G_0, or elasticity modulus at 1 Hz, and thereby the corresponding elasticity modulus $E \approx 3\,G_0$ (assuming an incompressible cellular medium) from:

$$G_0 = \frac{(2\pi)^\alpha}{A_0\,\Gamma(1+\alpha)} \tag{4.7}$$

where $\Gamma(1+\alpha) = \left[\int_0^{+\infty} \exp(-x)\, x^\alpha dx\right]$ is the Gamma Euler function.

Equations 4.4 through 4.6 represent the most frequent rheological models used to fit living cell data. An example of application obtained in Human Pulmonary Microvascular Endothelial Cells (HPMECs) is shown in Figure 4.2.

The single-Voigt model was used in early MTC experiments (Wang et al., 1993) and remains of interest because it assesses the CSK structure as a tensegrity structure, whose "solid-like" properties have already been demonstrated (Wendling et al., 1999). A unique viscoelastic component has also been useful to demonstrate the CSK specificity of the MTC technique using a drug (e.g., the cytochalasin D which disrupts the actin CSK) leading to a drastic (~50%) decrease in cell stiffness (Wang et al., 1993). The fundamental role of transmembrane integrin receptors in mechanotransduction across the cell membrane was demonstrated using the same single-Voigt model. To do so, the cell stiffness measured in living capillary endothelial cells using beads coated with integrin ligand (RGD) was compared to that obtained from control transmembrane metabolic acetylated low-density lipoprotein (AcLDL) receptors (Wang et al., 1993; Wang and Ingber, 1994).

The double-Voigt model was proposed to distinguish two major CSK structures, that is, the cortical and the deep CSK components, which exhibit a soft cortical CSK component compared to the stiff deep CSK component (Laurent et al., 2003). Such biological specificity of each component was demonstrated by treating or not tissue cells with cytochalasin D. The slow CSK compartment revealed the highest sensitivity to cytochalasin D treatment, consistent with the presumed highly polymerized nature of the deep CSK.

The power law model takes into account the fact that the intracellular medium exhibits a unique and weak power law behavior if the frequency range is limited (e.g., 5 Hz in Maksym et al., 2000 study; 10^{-1}–10^2 Hz in Trepat et al., 2004; 10^{-2}–10^2 Hz in Fabry et al., 2003). It has already been shown that the creep function may be described by a weak power law of elapsed time (Desprat et al., 2005; Lenormand et al., 2007). These authors have considered that cells behave like out-of-equilibrium and disordered systems in which multiscale structural rearrangements dominate through dynamic cross-linking, actin polymerization, and molecular motor activity. There are numerous examples in the literature of complex viscoelastic systems showing that rheological behaviors are timescale invariant, for example, colloidal systems close to sol/gel transition (Ponton et al., 2002), and "soft glassy materials" including foams, pastes, and emulsions (Weeks et al., 2000). According to these approaches, phenomena are supposed to occur at a multitude of length scales (i.e., from the individual actin filament size to actin bundles or stress fibers organized at cell size). Thus, characteristic time constants are multiple τ_i as if the cell structure was composed of an infinite series of Voigt elements with self-similar structures (Balland et al., 2006). These structures would be distributed according to the ideal power law: $\tau_i = \tau_m \, i^{\left(\frac{-1}{1-\alpha}\right)}$ (where τ_m represents the largest relaxation time in the cell), and the resulting creep function follows a weak power law of time. Although a series of elementary Voigt units may appear oversimplified to take into account active molecular mechanisms or filament remodeling, the various relaxation times associated with each viscoelastic unit can be considered to represent the storage and dissipation of elastic energy in actin filaments, bundles, and stress fibers. Dissipation itself may include several processes, such as CSK remodeling, molecular motor activity, and passive viscosity (Balland et al., 2006). Compared to single or double element viscoelastic solid-like models treated above, it appears that power law rheology is susceptible, due to its wide spectrum of characteristic time and length, to reveal some of the hidden molecular aspects which could not be captured by the two elementary viscoelastic models above.

Stamenovic et al. (2007) demonstrated that there is no need to invoke structural disorder/metastability to exhibit anomalous rheology in living cells. Using a broader range of frequencies, that is, 10^{-3}–10^4 Hz, they found that CSK dynamics is not timescale invariant but exhibits two power laws with a transition zone that can be explained by the nonequilibrium-to-near-equilibrium transition between protein–protein noncovalent interactions (Chowdhury et al., 2008). Based on microrheological models, it appears that MTC measurements on living cells most likely contain key molecular information but the challenge remains of how to capture the hidden biological information and how to relate the new parameters obtained to the cellular/molecular structure.

4.3.4 Assessment of Mechanical and Adhesion Properties from Transient MTC Loading

The imperfect recovery of the cellular deformation after transient MTC loading $\theta_{NR}(t)$ represents a significant proportion of the measured bead rotation $\theta(t)$ (usually between 10% and 30% of $\theta(t)$, depending on cell type and experimental conditions, see Equation 4.8). Nonreversible phenomena have been largely underconsidered in the past (see, for instance, the review on cell mechanics tools by Lele et al., 2007). By contrast, the recoverable component of cellular deformation $\theta_R(t)$ contains information on the mechanical response of twisted cells from which mechanical properties can be extracted.

The basic idea of the present approach is to consider that the nonrecovered bead rotation component $\theta_{NR}(t)$ contains key information on cellular adhesion that remains to be explored (Figure 4.2). To do so, we need to review molecular adhesion models. Previously, the nonrecovered bead rotation component $\theta_{NR}(t)$ was attributed to some cellular remodeling process that was supposed to be independent from the mechanical process. Indeed, the so-called "permanent deformation" reported during MTC (Wang, 1998; Wang and Ingber, 1994) was actually related to postloading cellular damage. By contrast, we presently consider that the nonrecovered bead rotation component $\theta_{NR}(t)$ actually reflects two phenomena that are related to cellular adhesion (see Equation 4.9): (1) a time-dependent sliding effect represented here by a sliding angular component $\theta_S(t)$ which is due to partial de-adhesion occurring primarily at the onset of loading and, to a lesser extent, during the loading application and (2) a time-independent sliding component θ_S^0, reflecting the few percent of surface adhesion sites that instantaneously dissociate at the beginning of loading because they failed to form effective attachments with the CSK. Therefore,

$$\theta(t) = \theta_R(t) + \theta_{NR}(t) \tag{4.8}$$

$$\theta_{NR}(t) = \theta_S(t) + \theta_S^0 \tag{4.9}$$

Note, however, that $\theta(t)$ is calculated from a modified Equation 4.2, that is, Equation 4.10, which takes into account the immediate sliding component θ_S^0 associated to a proportion "x" of free rotating beads (Fabry et al., 1999):

$$\theta(t) - \theta_S^0 = \text{arc} \cos\left(\frac{M(t)}{M_0(1-x)}\right) \tag{4.10}$$

"100 x" can be seen as the percentage of slip bonds present in a given cell culture (usually <5%).

The time-dependent signal of MTC can be indifferently represented in terms of bead deviation angle $\theta(t)$ (Equation 4.10) or in terms of remanent magnetic moment $M(t)$ (as in Figure 4.2) using the reciprocal expression of Equation 4.10:

$$M(t) = M_0(1-x)\cos\left(\theta(t) - \theta_S^0\right) \tag{4.11}$$

The solution for the three viscoelastic solid-like models is integrated in a more general model, which takes into account mechanical and adhesion properties from the time-dependent recoverable and unrecoverable components of the MTC signal.

For the viscoelastic solid-like model, derived from a single-Voigt model:

$$\theta_R(t) = \frac{\sigma}{E}\left(1 - \exp\left(-\frac{t}{\tau}\right)\right) \tag{4.12}$$

For the viscoelastic solid-like model, derived from a double-Voigt model:

$$\theta_R(t) = \frac{\sigma}{E_1}\left(1 - \exp\left(-\frac{t}{\tau_1}\right)\right) + \frac{\sigma}{E_2}\left(1 - \exp\left(-\frac{t}{\tau_2}\right)\right) \tag{4.13}$$

For the time-invariant power law model:

$$\theta_R(t) = \sigma\, A_0\left(\frac{t}{t_0}\right)^\alpha \tag{4.14}$$

Figure 4.2 illustrates the validity of these three rheological models in a culture of living tissue cells (HPMVEC).

4.3.5 Molecular Adhesion Models

The strength of molecular links created between the beads and the cell is of particular interest for the MTC technique because these interfacial links (e.g., the transmembrane integrin receptors bound to the synthetic

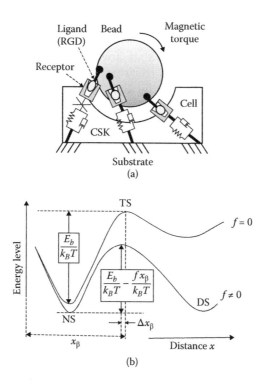

FIGURE 4.3

The effect of force at cellular and molecular scales (a) The effect of twisting forces on bound receptors and CSK. The mechanical properties of one or several viscoelastic solid-like models are measured through integrins whose partial detachment allows characterizing the kinetics of de-adhesion using an empirical model at the cell scale and a theoretical model down to the molecular level of the bead–cell interaction. (b) The effect of force on the energy-distance landscape for a single or collective receptor–ligand bond(s). Depending on the separation distance between two macromolecules which chemically interact, the energy level changes, passing from native state (NS) to transition state (TS) located in between bond association and dissociation. The strength of a bond is classically given by the maximum gradient $-(\partial E_b/\partial x)_{max}$ of the interaction potential defined along the direction of separation x. When a force is exerted on the bond, the height of the energy barrier is lowered by a quantity given by mechanical energy, producing an increase in dissociation rate more rapidly leading to the dissociation state DS. We assume that the distance x_β at which TS is observed is only slightly modified by force application, that is, $\Delta x_\beta \approx 0$.

RGD-ligand covering the beads) transmit to the cell structure the mechanical loading generated by MTC (Wang et al., 1993) (Figure 4.3a). A general concept is to consider that the strength of molecular attachment to a cell membrane receptor is determined by the weakest component of the complex, that is, the ligand–receptor bond or anchorage of the receptor to the membrane (Evans and Ritchie, 1994). We propose to apply this concept to MTC measurements. It is worth noting that the step loading signal used in standard MTC experiments is a succession of force-ramp and force-clamp maneuvers (Figure 4.1b), now the reference method for the characterization of "weak" interfacial interactions (Evans and Kinoshita, 2007). For similar reasons, we can assume that any defect in bead–cell adhesion would be responsible for partial detachment of preexisting bonds created after a short period of bead incubation. In the improved version of the MTC technique presently described, we postulate that the amount of sliding measurable between the bead and the cell after a first twist could contain some key information on the kinetics of bead–cell attachment and thus reveal some hidden aspects of cellular adhesion. This "sliding effect" has often been reported in micromanipulation experiments in which transient loadings are applied but its analysis went never beyond a hypothetical plastic permanent deformation of undetermined origin (Lele et al., 2007; Wang, 1998). Our goal is to apply an up-to-date theory of bond dissociation to reveal some hidden aspects of the cellular response to transient loadings and describe, by appropriate models, the kinetic parameters of the receptor–ligand unbinding (Isabey et al., 2016) (Figure 4.3).

Molecular adhesion at surfaces is governed by essentially reversible, non-covalent interactions between large molecules (Evans, 2001). Unlike intimate covalent bonds, which are strong intramolecular bonds, noncovalent intermolecular bonds are weak by nature and thus have limited lifetimes (Evans, 1998). They actually must work together to have a significant effect, combined bond strength being much greater than the sum of individual bonds (Evans, 2001; Isabey et al., 2013; Williams, 2003). Dynamic force spectroscopy is the reference method allowing the experimental characterization of the complexity of molecular interactions between receptor and ligand bonds (Evans and Ritchie, 1997). For instance, by measuring the rupture force at different force regimes, the various barriers of the potential energy profile can be revealed (Evans, 1998). In the isolated bond under zero force, thermal fluctuations cause spontaneous dissociation and reassociation after a time t_{off} representing the natural lifetime which may widely vary, for example, from a few seconds in selectin bonds (Evans et al., 2001; Fritz et al., 1998) to a few hours for antigen–antibody interaction (Hinterdorfer et al., 1996; Schwesinger et al., 2000), and even several days for biotin–streptavidin (Chilkoti and Stayton, 1995). In other words, for timescales above t_{off}, external force is not needed to break receptor–ligand bond because the low thermal energy level $k_B T$ (≈ 4.1 pN·nm) is sufficient to overcome the energy barrier (E_b) which separates the bound state from the unbound state (Figure 4.3b; k_B is the Boltzmann constant and T is the temperature). Spontaneous dissociation of a bond can be characterized by the dissociation rate at zero force k_{off}^0 issued from the diffusion theory of a Brownian particle across an energetic barrier in a viscous environment (Hänngi et al., 1990; Kramers, 1940).

$$k_{\mathrm{off}}^0 = \frac{1}{t_{\mathrm{off}}^0} = \frac{1}{t_D} \cdot \exp\left(\frac{-E_b}{k_B T}\right) \tag{4.15}$$

E_b is the intermolecular chemical potential or initial barrier height. The exponential dependence of the dissociation rate at zero force results in huge variations in the kinetics of dissociation which in liquids starts from the highly elevated dissociation frequencies of $t_D^{-1} \sim 10^9$–10^{10} s^{-1} (t_D being the ultrafast Brownian diffusion time) at $E_b = 1$ $k_B T$ to decrease at $t_D^{-1} \sim 10^0$ s^{-1} at $E_b = 21$ $k_B T$ and even the incredibly slow $t_D^{-1} \sim 1/(40$ years$)$ for elevated barriers of $E_b = 42$ $k_B T$ (Evans, 2001).

Bonds stretched with forces resist detachment on timescales below $t_{off}^0 \left(= 1/k_{off}^0 \right)$. The rupture forces, which reflect the thermally activated kinetics, can be classically obtained by ultrasensitive probes with high-resolution detection such as AFM (Florin et al., 1994). The intuitive effect of force on bonds is to lower the barriers of the energy landscape (see Figure 4.3b), thus facilitating the transition (TS) from the bond state to the unbound state as described by the single-bond lifetime decrease, earlier predicted by Bell (1978):

$$k_{off}(f) = k_{off}^0 \cdot \exp\left(\frac{f}{f_\beta}\right) \qquad (4.16)$$

f_β was a characteristic force related to thermal energy. Assuming that the kinetic transition is dominated by the shape and height of the energy barrier, Evans and Ritchie have proposed a generic expression of the dissociation rate under force with a solid physical background based on Kramer's theory (Evans and Ritchie, 1997):

$$k_{off}(f) = t_{off}^{-1}(f) = k_{off}^0 \cdot g\left(\frac{f \cdot x_\beta}{k_B T}\right) \cdot \exp\left(\frac{f \cdot x_\beta}{k_B T}\right) \qquad (4.17)$$

$f \cdot x_\beta$ represents the mechanical energy calculated in the direction of the applied force for an interaction distance x_β at which the transition state is presumably observed (see Figure 4.3b). This disjoining potential represents the reduction in height of the energy barrier. The distance x_β provides a reference force scale f_β:

$$\frac{k_B T}{x_\beta} = f_\beta \qquad (4.18)$$

f_β represents the amount of force which must be applied to a bond in order to decrease the energy barrier by $k_B T$, thus producing an increase in the dissociation rate by a factor 2.7. Typical values of these reference parameters are ≤ 1 nm ($= 10^{-9}$ m) for x_β and 10 pN for f_β (1 pN $= 10^{-12}$ N), for example, for an integrin–VCAM bond at the biological temperatures: $x_\beta \approx 0.30$ nm, $f_\beta \approx 13$ pN,

$k_{off}^0 \approx 0.8$ s^{-1} (Evans and Kinoshita, 2007). The function $g(f/f_\beta)$ takes into account the force-induced modification of the shape of the energy landscape. For sharp energy barriers, forces application is supposed not to affect the shape of the energy barrier and the position of the kinetic transition, therefore $g(f/f_\beta) \approx 1$ (Evans and Ritchie, 1997). In such a case, the distance between transition states with and without force applied is negligible, that is, $\Delta x_\beta \approx 0$ (Figure 4.3b). For more complex energy barriers, the dimensionless function: $g(f/f_\beta)$ determines the force-driven amplification of the kinetic properties of the bond. For noncooperative bonds working as a zipper, $g(f/f_\beta) \approx \dfrac{1}{N}$, meaning that g is independent of the force but is inversely proportional to the number of zipper bonds (Isabey et al., 2013; Williams, 2003). The leading bond receives the totality of the force applied until failure, then force is transferred to the next bond and so on. While the force scale of the collective zipper bond is unchanged compared to the force scale of the single bond, the dissociation rate of N noncooperative bonds in a zipper-like configuration is N times smaller than the dissociation rate of a single bond.

During bead twisting experiments (MTC), shear deformation dominates and remains at the μm-scale at or near the bead surface (Ohayon et al., 2004), meaning that the zipper bond configuration seems the most appropriate. Moreover, we can neglect force rebinding because displacements are significant and, far from equilibrium (NS in Figure 4.3b), the rate of rebinding decreases exponentially and tends toward zero (Williams, 2003).

To describe the unbinding phase during MTC experiments, we propose to use a generalized form of Bell's equation, which is still consistent with Kramer's results:

$$K_{off}(f) = K_{off}^0 \cdot \exp\left(\frac{f}{F_\beta}\right) \text{ with } K_{off}^0 = \frac{k_{off}^0}{N} \text{ and } F_\beta \approx f_\beta \text{ for zipper bonds} \quad (4.19)$$

Equation 4.19 summarizes the most common understanding previously described by Equations 4.16 and 4.17, which is to consider that mechanical force applied to a bond tends to shorten its lifetime. Similarly, the form of the dissociation rate given by Equation 4.19 reflects the solution of a first-order kinetic equation, which describes the probability $P(t)$ that a collective bond will survive in the bound state:

$$\frac{dP(t)}{dt} = -K_{off}(f) \cdot P(t) \text{ with } P(t) = \exp\left\{-\int_0^t K_{off}(f) \; dt'\right\} \quad (4.20)$$

It is worth noting that during MTC experiments, the probability to survive $P(t)$ can be simply related to the time-dependent sliding angle $\theta_S(t)$ by using the most straightforward function varying between $P(t)=0$ (for $\theta_S(t)=0$) and $P(t)=1$ (for $\theta_S(t)=\pi/2$), namely the cosine function:

$$P(t) = \cos(\theta_S(t)) \tag{4.21}$$

$P(t)$ is then used to calculate K_{off}^0 and F_β (Equation 4.19) using best curve fitting between experimental signal and the elementary models of time-dependent probability described below. An example of the appropriateness of the proposed approach is shown in Figure 4.2.

As previously suggested (Evans and Kinoshita, 2007), we can consider two elementary cases of force dependence which advantageously correspond to the actual experimental conditions used during MTC (Figure 4.4): (1) an almost constant force loading during 60 seconds of magnetic torque application (called force-clamp period), between two (2) linear time-dependent force-loading (~400 ms) (called force-ramp) periods. Importantly, these two particular loadings enable deducing the dissociation rate at zero force K_{off}^0 and a reference force F_β.

In the force-clamp period, the probability of bond survival exponentially decays with time t which requires a unique (i.e., time-independent) dissociation rate to be characterized:

$$P(t) = \exp\left[-K_{off}(f_{cste}) \cdot t\right] = \exp\left[-K_{off}^0 \cdot \exp\left(\frac{f_{cste}}{F_\beta}\right) \cdot t\right] \tag{4.22}$$

Equation 4.22 indicates that by measuring the statistics of bond failure at different force levels, that is, the ratio between the number of bonds $\dfrac{N(t)}{N_{tot}}$ existing at a given instant and the number of bonds before torque application, it is possible to assess the mean dissociation rate or mean lifetime, $T_{off}(f = \text{cste}) = \dfrac{1}{K_{off}(f = \text{cste})}$, from the (constant) slope of the above relationship (Evans and Kinoshita, 2007). During MTC experiments, the unrecoverable component of angular bead deviation, called the "sliding angle," can be estimated and related to the statistics of bond detachment (see Figure 4.4) during force clamp but also during the force ramp as shown below.

In the force-ramp periods, that is, $df/dt = r_f = \text{cste}$, the probability of bond survival (or distribution of lifetime) is no longer a single-exponential decay function with time but reveals a loading rate-dependent decay such that its timescale is dramatically reduced with each order of magnitude increase in ramp rate (Evans and Ritchie, 1997; Thomas et al., 2008).

$$P(t) = \exp\left[-\frac{K_{off}^0 \cdot F_\beta}{r_f}\left(\exp\left(\frac{r_f\, t}{F_\beta}\right) - 1\right)\right] \tag{4.23}$$

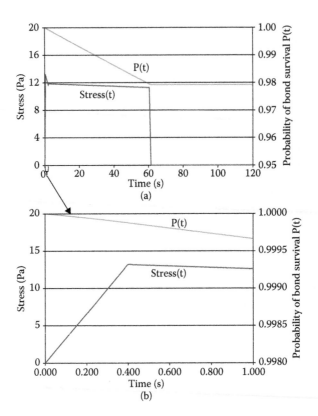

FIGURE 4.4
Time course of the twisting stress (in Pa, left vertical scale) and probability of bond survival
$P(t)$ (right vertical scale) in living HPMECs are presented: (a) For the entire cell response, that
is, over the 120 seconds duration of the experiment. (b) For the onset of cell response, that is,
over the first second of loading. The actual stress signal (in Pa) resembles to the generated
signal in Figure 4.1b with, however, a slight nonsignificant decay caused by the decrease in the
angle between $\vec{m}$ and $\vec{H}$. The number of bonds decreases as a sustained stress of 60 seconds
is applied. Values of $P(t)$ show that less than 3% of bonds dissociate, which is yet sufficient to
calculate the kinetics parameter of de-adhesion.

A characteristic scale for ramp rate $r_f^0 \left(= K_{off}^0 \cdot F_\beta\right)$ can be deduced from
kinetic parameters governing bond dissociation (Equation 4.19; r_f^0 defines
the loading speed above which the bond, driven by force "far from equilibrium," dissociates much faster than its apparent lifetime T_{off}^0 and thus resists
force). Using Equation 4.20, it is possible to calculate the most probable dissociation rate K_{off}^* from the maximum value of the time derivative of $P(t)$
(Equation 4.23). This value is actually related to the most likely rupture force
F^* by the following expression (Evans and Calderwood, 2007):

$$\frac{1}{K_{off}^*} = \frac{F^*}{r_f} \left(= T_{off}^*\right) \tag{4.24}$$

T_{off}^* is the most probable lifetime. So, the likelihood of bond lifetime T_{off}^* and the likelihood of rupture forces F^* are closely related through the kinetic theory of bond failure. The two parameters consistently depend on r_f, K_{off}^0, and F_β.

During MTC experiments, the most probable lifetime or rupture force would not be attained because the maximal value of the derivative of $P(t)$ is never attained within the 60 seconds of the experiment. However, subcritical levels of mean rupture force are sufficient to access to a partial but significant range of lifetime distribution, which enables us to deduce some key parameters of adhesion kinetics, namely K_{off}^0 and F_β. Note that if r_f could not easily be changed by order of magnitude during MTC experiments, other parameters such as N, K_{off}^0, and F_β could easily vary with experimental conditions.

4.3.6 MTC Application to Characterization of a Living Cell Response

To illustrate the capability of the MTC technique to characterize the short- and long-term responses of stressed living cells, we present below the results obtained in HPMECs, stimulated or not by sustained twisting of RGD-coated beads (Wang et al., 2012). In the pioneering study of Wang et al. (1993), capillary endothelial cells plated on fibronectin-coated bacteriological plastic were used to demonstrate the advantages of RGD-coated magnetic beads to assay both mechanotransduction pathways and mechanical properties of the CSK. Since that time, the MTC technique has enabled the testing of different types of transmembrane adhesion receptors, including various types of integrins, cadherins, selectins, ICAM-1, and urokinase receptors, looking at their ability to support transmembrane mechanical coupling to the CSK (Planus et al., 2005; Potard et al., 1997; Wang and Ingber, 1995; Yoshida et al., 1996). A key criterion in testing surface receptors with MTC is the link with the CSK permitted by RGD ligands or anti-integrin antibodies, while transmembrane receptors which do not physically link the internal CSK such as metabolic receptors, growth factor receptors, or histocompatibility antigens failed to exhibit resistance to angular rotation and thereby measured negligible stiffness (Wang et al., 1993; Yoshida et al., 1996). Another way to demonstrate the specific role of the CSK in the cellular response to bead twisting is to use pharmacological drugs or genetic knock-out techniques that specifically alter one or several of the three cytoskeletal filament networks, that is, actin filaments, microtubules, and intermediate filaments (Wang, 1998; Wang et al., 1993). These results have also shown that the actin filament network was found to play a major role in the cell's resistance to stress. Coupling MTC with signal transduction experiments enabled us to reveal new molecular pathways of cellular mechanotransduction, for example, the mechanical activation of gene transcription driven by the alteration of the cyclic AMP signaling cascade (Meyer et al., 2000) or the regulation of endothelin-1 gene expression in a prestress-dependent manner (Chen et al., 2001). In spite of the strength of MTC to illuminate such sophisticated molecular features germane to cell function, kinetic adhesion parameters have never been explicitly derived from MTC experiments in spite of the idea that MTC was initially proposed to assess ligand–receptor and/or receptor-CSK strengths (Wang et al., 1993) (Figure 4.5).

Figure 4.5 illustrates the new knowledge gained by the MTC technique while both mechanical and adhesion properties are extracted from the living cell response signal, that is, HPMECs mechanically unstressed or stressed during 2 hours of sustained, cyclic twisting. We specifically compared the results obtained with the viscoelastic solid-like model and a double-Voigt component combined with empirical (Figure 4.5a, c, and e) or theoretical (Figure 4.5b, d, and f) models of de-adhesion. The empirically based model of de-adhesion considers the partial sliding between bead and cell, that is, de-adhesion process at cell scale, while the theoretically based model integrates adhesion kinetics, that is, de-adhesion process at molecular scale.

The main messages that can be drawn from these MTC results are summarized as follows:

> The three microrheological models tested (see paragraphs 3, 4, and Figure 4.2) combined either with the empirical or theoretical de-adhesion models (see paragraphs 4, 5, and Figure 4.4) provide a satisfactorily curve fitting to the raw MTC data with correlation coefficients above 0.95. This means that the cell response to transient bead twisting is not exclusive of a given model and contains a rich panel of biomechanical information including mechanical properties of the CSK and kinetic parameters of surface molecular linkages at the cell surface.
>
> The cortical (Figure 4.5a and b) and deep (Figure 4.5c and d) CSK components exhibit viscoelastic time constants that differ by orders of magnitude. The viscoelastic response of the cortical CSK component is, as expected (Laurent et al., 2003), faster than the deep CSK component. Both time constants consistently fall in the experimental range of the transient MTC signal within 60 seconds or less. The longer relaxation time constants characterizing the deep CSK component suggest characteristic elements of longer size by contrast to the shorter relaxation time constants that can be associated with the short interconnected filaments characterizing the cortical CSK (Balland et al., 2006).
>
> Stimulated cells exhibit highly significant decrease in τ_1 (Figure 4.5a and b) which is essentially due to a significant decrease in friction modulus (results not presented). These changes in CSK viscoelastic properties are consistent and quantify the remodeling of local actin network in response to sustained mechanical stimulation (Wang

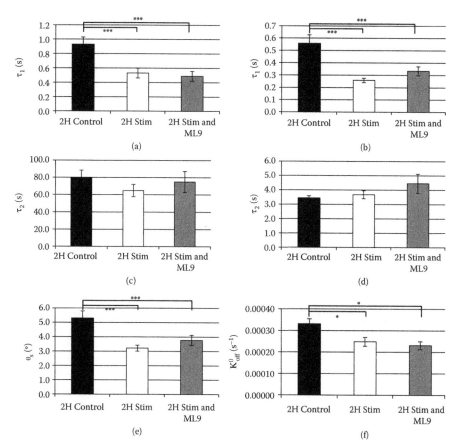

FIGURE 4.5

Mechanical properties and kinetic adhesion parameters measured by MTC in living HPMECs. Graphs on the left (a, c, and e) correspond to the empirical cell-scale model of de-adhesion while graphs on the right (b, d, and f) correspond to the theoretical molecular-scale model of de-adhesion. A unique viscoelastic solid-like model is presented herein: the double-Voigt model which accounts for the cortical and the deep CSK components. (a) Viscoelastic time constant of the cortical CSK component τ_1 (in seconds) obtained with the double-Voigt viscoelastic model coupled with the empirical model of de-adhesion. (b) Viscoelastic time constant of the cortical CSK component τ_1 (in seconds) obtained with the double-Voigt viscoelastic model coupled with the theoretical model of de-adhesion. (c) Viscoelastic time constant of the deep CSK component τ_2 (in seconds) obtained with the double-Voigt viscoelastic model coupled with the empirical model of de-adhesion. (d) Viscoelastic time constant of the deep CSK component τ_2 (in seconds) obtained with the double-Voigt viscoelastic model coupled with the theoretical model of de-adhesion. (e) Maximum value of the time-dependent sliding angle $\theta(t)$ (in °) measured at the end of the initial ramp of stress and obtained with the double-Voigt viscoelastic model coupled with the empirical model of de-adhesion. (f) Dissociation rate (in s^{-1}) at zero force obtained with the double-Voigt viscoelastic model coupled with the theoretical model of de-adhesion. Mechanical stimulation of HPMECs cultured in monolayer is obtained by sustained cyclic twisting of the ferromagnetic beads for 2 hours (around 10 Pa at 3 Hz), the cell culture being placed inside an incubator. In some experiments, HPMECs were pretreated by myosin light chain kinase inhibitor (ML9) (at 50 µM) before exposure to mechanical stimulation. Horizontal bars indicate statistically significant differences (*$p \leq .05$; **$p \leq .01$; ***$p \leq .001$). Vertical error bars are SEM.

et al., 2012). The relative constancy of τ_2 (Figure 4.5c and d) suggests a weaker sensitivity of this parameter to CSK remodeling but separate evolutions of both viscous and elasticity modulus suggested that CSK remodeling also impact the deep CSK compartment (presently not shown). Treatment by an actomyosin blocker such as ML9, an inhibitor of myosin light chain kinase (MLCK), is not sufficient to counterbalance the effect of mechanical stimulation on the evolution of CSK mechanical properties. MLCK has already been found to play an important role in chemically induced endothelial dysfunction but this signaling pathway is not the unique pathway controlling actomyosin activation (Tinsley et al., 2000).

We observed satisfactory curve fitting with the time-invariant power law model (presently tested in combination with the adhesion model but results are not presented herein). This model led to rather low values of the exponent $\alpha (= [0.12 - 0.08])$ which is consistent with the weak power dependence characterizing living cells (Chowdhury et al., 2008; Fabry et al., 2003; Puig-de-Morales et al., 2004). Such values are at the bottom of the range of α-values currently measured in a variety of living cells, that is, $\theta (= [0.15 - 0.35])$ (Alcaraz et al., 2003; Balland et al., 2006; Féréol et al., 2008; Trepat et al., 2004). These results confirm that living cells stay mostly on the solid-like side of the transition solid-fluid (Fabry et al., 2003). Importantly, the power law model (Equation 4.6), which associates a large number of elementary solid Voigt units in series, and therefore covers a wide range of timescales of the cytoskeletal network (Balland et al., 2006), might account for a part of the molecular phenomena that are otherwise taken into account by the adhesion model (Equations 4.14 and 4.21). The power law model and the adhesion model could well describe molecular phenomena with some redundancy. By contrast, longer viscoelastic time constants, well taken into account by the two elementary solid-like models used in Figure 4.5, could describe phenomena close to the cell scale, that is, by viscoelastic models with a limited number of time constants.

The maximum value of the sliding angular component, θ_S, is normally attained at the instant of maximum stress, that is, right after the initial ramp of loading (Figure 4.4). This parameter reflects the maximal state of de-adhesion, which is significantly decreased after mechanical stimulation by the empirical cell-scale model (Figure 4.5e). The theoretical model at molecular scale shown in Figure 4.5f reveals a significant decrease in K_{off}^0, that is, roughly from 0.0003 to 0.0002 s^{-1}, which suggests that the number of noncooperative integrins acting at the cell–bead interface may have been doubled in response to mechanical stimulation (Equation 4.19). Incidentally, the reference force F_β was not deeply changed by mechanical stimulation suggesting that the shape of the overall energy function was not strongly modified after stimulation. Such change in K_{off}^0 from nonstimulated to stimulated HPMECs appears consistent with biological observations performed in the same cell culture after stimulation by mechanical stress (Wang et al., 2012). Indeed, integrin staining, and specifically $\alpha_v\beta_3$integrins, revealed clustering and relocalization of these adhesion molecules at the cell–matrix interface, that is, near the basal face. We can consider that the bead is "sensed" by the cell as a localized extracellular matrix and, thereby, the cell attachment to the matrix coating the bead could most likely simulate the overall cellular response to extracellular matrix.

Comparing the control (respectively, mechanically stimulated) values of K_{off}^0 (= 0.0003 (respectively, 0.0002) s^{-1}) with the unstressed dissociation rates reported in the literature for single bonds implicating activated integrin, e.g., k_{off}^0~0.012 s^{-1} for fibronectin binding (Li et al., 2003) the average number of integrins effectively linked to the bead ligand would be in the range of N = 40–60, assuming zipper bond configuration. For a bead surface whose maximum area is 65 μm^2, previous considerations lead to a bond density of only a few integrins/μm^2 which is reasonable assumption on the apical cell surface.

On the whole, the significant decrease in viscoelastic time constants (Fig. 5 A, B) is consistent with an increase in intracellular tension in HPMECs (Wang et al., 2002) that would be induced by mechanical stimulation. Cell stimulation would result in reinforcement of cell–matrix interactions which in turn affects cell–cell junctions and notably the cadherin–cadherin CSK-related bonds. The functional consequence is an increase in permeability of the HPMEC monolayer suggested by an increase in the leakage measured by a fluorescent marker (dextran) (Wang et al., 2002).

4.3.7 Some Remarkable Features of MTC

The simultaneous characterization of mechanical and adhesive properties in living cells is rendered possible during MTC because cells create firm attachment to the beads while the bead–cell system fails to recover its initial position (Isabey et al., 2016). The firm CSK attachment has been largely demonstrated since the very first MTC studies (Wang et al., 1993; Wang and Ingber, 1994) and bead coating for mechanoreceptors physically related to the CSK constitutes a key feature of the MTC method. By contrast, the partial remodeling of the bead–cell attachment during twisting has been largely ignored and if considered, uniquely attributed to a guessed plastic CSK deformation with no relation to interfacial molecular structure (Lele et al., 2007).

Two biological factors are important for bead attachment to the CSK through actin CSK formation and strengthening of bead attachment permitting, for instance, rearward translocation: (1) integrin assembly called "clustering" of the integrins and (2) ligand binding (Choquet et al., 1997; Felsenfeld et al., 1996). Translocation of beads coated with FN7-10 at first form a relatively weak attachment to the CSK, and they can be pulled off by a modest force using a laser trap (Choquet et al., 1997). Such a situation is reflected by the beads experimenting immediate sliding as quantified in the present approach by the component θ_S^0 associated to a percentage "x" of free rotating beads (Equation 4.10). Interestingly, this percentage tends to decrease after stimulation (results not shown). However, when beads were subjected to a sustained force of 10 seconds duration or more, the cell forms actin CSK upon bead contact and increases the bond strength (Choquet et al., 1997). Such reinforcement most likely occurs in the course of 60 seconds of MTC loading which most likely contributes to maintain a firm CSK attachment during MTC experiment. This attachment is clearly enhanced after 2 hours of cyclic twisting while the same beads are used for both MTC measure and sustained mechanical stimulation as shown in HPMECs (Wang et al., 2012). CSK recruitment may also be related to the findings that when FN-coated beads are placed in contact with a cell, they induced formation of an actin CSK immediately underneath the point of bead binding (Miyamoto et al., 1995a,b).

Another remarkable feature of MTC when applied to HPMECs is that the effective stress exerted by the bead to the HPMEC cells, that is, 9–24 Pa (1 Pa = 1 pN/μm^2), remains well below the physiological range of maximal stress, for example, ~1,000–2,000 Pa, that a cell can exert through a given focal adhesion site (Munevar et al., 2001; Oliver et al., 1999). Therefore, each integrin whose number—according to present results—might not exceed a few integrins per μm^2 would contribute to a reasonable value of applied force namely not exceeding a few tens of pN during MTC, which corresponds to forces generated by about ten myosin II motors interacting with a single actin filament (Ishijima et al., 1994; Molloy et al., 1995). These estimates of the force generated per bond or of the dissociation rate belong to the deterministic approach which consists of finding mean values of parameters while small-scale phenomena are probabilistic and thus lack definitive values (Zhu et al., 2000). Due to the stochastic nature of receptor–ligand bonds, experiments with small number of bonds per cell must be treated by probabilistic theory for kinetics of small systems (Cozens-Roberts et al., 1990). An averaging procedure is actually integrated in the measurement of the remanent magnetic field, which reflects the entire bead and cell populations. MTC analysis works as if cellular and molecular behaviors were assessed from the averaged behavior of one bead whose motion is representative of the mean behavior of around 100,000 beads and 50,000 cells in each culture. Thus, in principle, parameters measured by MTC are the result of a homogenization that provides a deterministic estimate of definitely stochastic mechanisms (Ohayon et al., 2004).

Another important feature of the cell response to bead twisting is the stress or strain hardening observed mainly in tissue cells and to a lesser extent in less tensed inflammatory cells (Féréol et al., 2008). Bead twisting resulted in significant cell surface deformation (i.e., 15°–60° corresponding to 500–2,500 nm surface displacements) which, in highly structured tissue cells, produced long-distance deformations contributing to this reported stress-dependent increase in stiffness. Coupling MTC with traction force microscopy and MTC in contractile cells such as human airway smooth muscle (HASM) cells, Wang et al. found that cell stiffness is linearly related to prestress or intracellular tension generated by cytoskeletal actomyosin interactions (Wang et al., 2002). These stress related behaviors, that is, stress hardening and prestress, are both consistent with the prediction of cellular tensegrity models (Cañadas et al., 2006; Coughlin and Stamenovic, 1998; Stamenovic and Coughlin, 1999; Wendling et al., 2000). The tensegrity approach may also explain why the cortical CSK structure is less tensed than the deep CSK structure (Laurent et al., 2002a). The structural approach is consistent with the idea that, by applying torque through integrins, MTC probes the underlying 3D structure of the CSK rather than the 2D structure of the cell's surface membrane (Stamenovic and Coughlin, 2000; Wendling et al., 1999).

4.4 Conclusions and Perspectives

We have shown here that transient stress MTC contains key information on cell adhesion that can be obtained in addition to mechanical properties over a large cell population. MTC can be advantageously combined with other techniques, such as signal transduction experiments or gene transcription activation, in order to measure CSK mechanical properties or to study the role of cytoskeletal tension in parallel to the evaluation of specific mechano-transduction or signaling pathways (Chen et al., 2001; Hu et al., 2003; Meyer et al., 2000) or to study the cellular response to mechanical stimulation, as presently done.

A major evolution has been achieved by coupling MTC with microscopic visualization, usually made at the single cell level. This is usually done by fabricating a device containing orthogonal Helmholtz coils placed on a microscopic stage. In such a case, twisting forces are applied to many beads on a cell while using optical techniques enables local analysis of the structure and the biochemical response of beads located in specific positions on the cell surface or inside the cytoplasm. The coupling of MTC and visualization has been advantageously used (1) to characterize the frequency response with the so-called oscillatory MTC (Fabry et al., 2001) and (2) to explore nanoscale displacements of molecular elements and stress distribution inside the cytoplasm and the nucleus with the so-called 3D magnetic tomography (Hu et al., 2003).

Bead twisting has become a generic method, facilitating a wide array of applications in living cells cultured on 2D substrates. It is, however, important to mention that future applications of MTC could focus on investigations of cellular responses in 3D networks, for example, various collagens or gels of different concentrations and mechanical properties (Isabey et al., 2016). To our knowledge, such applications have not yet been explored, although there is no technical limitation to extend the remanent magnetic field measurements to 3D environments in which cells probed with beads are immersed. In vivo, the tissue is a network of three-dimensionally interconnected multivalent ECM ligands, predominantly consisting of a collagen fiber backbone interconnected with fibronectin, hyaluronan, and other components. The vast majority of cellular and subcellular mechanical studies have been performed on cells attached to 2D substrates, whereas less attention has been given to the more physiological, realistic 3D matrix environment. Furthermore, nearly all of these cell rheological studies have focused on gels with high stiffness values ignoring softer substrates while they represent, in a more realistic fashion, the extracellular in vivo condition. MTC applied to living cells in 3D environments could provide quantitative understanding of intracellular mechanics and interactions with complex and diverse matrices. These questions are also interesting regarding pathology, such as cancer cell invasion, since clinical and animal data (Paszek et al., 2005; Samani et al., 2007; Zaman et al., 2006) suggest a correlation between tissue density, cancer aggressiveness, and mechanical factors.

References

Alcaraz, J., Buscemi, L., Grabulosa, M., Trepat, X., Fabry, B., Farre, R., Navajas, D., 2003. Microrheology of human lung epithelial cells measured by atomic force microscopy. *Biophys J* 84, 2071–2079.

Balland, M., Desprat, N., Icard, D., Fereol, S., Asnacios, A., Browaeys, J., Henon, S., Gallet, F., 2006. Power laws in microrheology experiments on living cells: Comparative analysis and modeling. *Phys Rev E Stat Nonlin Soft Matter Phys* 74, 021911.

Bell, G.I., 1978. Models for the specific adhesion of cells to cells. *Science* 200, 618–627.

Brain, J.D., Bloom, S.B., Valberg, P.A., Gehr, P., 1984. Correlation between the behavior of magnetic iron oxide particles in the lungs of rabbits and phagocytosis. *Exp Lung Res* 6, 115–131.

Cai, S., Pestic-Dragovich, L., O'Donnell, M.E., Wang, N., Ingber, D., Elson, E., De Lanerolle, P., 1998. Regulation of cytoskeletal mechanics and cell growth by myosin light chain phosphorylation. *Am J Physiol* 275, C1349–C1356.

Cañadas, P., Wendling-Mansuy, S., Isabey, D., 2006. Frequency response of a viscoelastic tensegrity model: Structural rearrangement contribution to cell dynamics. *J Biomech Eng* 128, 487–495.

Chen, J., Fabry, B., Schiffrin, E.L., Wang, N., 2001. Twisting integrin receptors increases endothelin-1 gene expression in endothelial cells. *Am J Physiol Cell Physiol* 280, C1475–C1484.

Chicurel, M.E., Chen, C.S., Ingber, D.E., 1998. Cellular control lies in the balance of forces. *Curr Opin Cell Biol* 10, 232–239.

Chilkoti, A., Stayton, P.S., 1995. Molecular-origins of the slow streptavidin-biotin dissociation kinetics. *J Am Chem Soc* 117, 10622–10628.

Choquet, D., Felsenfeld, D.P., Sheetz, M.P., 1997. Extracellular matrix rigidity causes strengthening of integrin-cytoskeleton linkages. *Cell* 88, 39–48.

Chowdhury, F., Na, S., Collin, O., Tay, B., Li, F., Tanaka, T., Leckband, D.E., Wang, N., 2008. Is cell rheology governed by nonequilibrium-to-equilibrium transition of noncovalent bonds? *Biophys J* 95, 5719–5727.

Cohen, D., 1973. Ferromagnetic contamination in the lungs and other organs of the human body. *Science* 180, 745–748.

Coughlin, M.F., Stamenovic, D., 1998. A tensegrity model of the cytoskeleton in spread and round cells. *J Biomech Eng* 120, 770–777.

Cozens-Roberts, C., Lauffenburger, D.A., Quinn, J.A., 1990. Receptor-mediated cell attachment and detachment kinetics. I. Probabilistic model and analysis. *Biophys J* 58, 841–856.

Crick, F.H.C., Hughes, A.F.W., 1950. The physical properties of cytoplasm. *Exp Cell Res* 1, 37–80.

Desprat, N., Richert, A., Simeon, J., Asnacios, A., 2005. Creep function of a single living cell. *Biophys J* 88, 2224–2233.

Evans, E., 1998. Energy landscapes of biomolecular adhesion and receptor anchoring at interfaces explored with dynamic force spectroscopy. *Faraday Discuss* (111), 1–16.

Evans, E., 2001. Probing the relation between force--lifetime--and chemistry in single molecular bonds. *Annu Rev Biophys Biomol Struct* 30, 105–128.

Evans, E.A., Calderwood, D.A., 2007. Forces and bond dynamics in cell adhesion. *Science* 316, 1148–1153.

Evans, E., Kinoshita, K., 2007. Using force to probe single-molecule receptor-cytoskeletal anchoring beneath the surface of a living cell. *Methods Cell Biol* 83, 373–396.

Evans, E., Leung, A., Hammer, D., Simon, S., 2001. Chemically distinct transition states govern rapid dissociation of single L-selectin bonds under force. *Proc Natl Acad Sci USA* 98, 3784–3789.

Evans, E., Ritchie, K., 1994. Probing molecular attachments to cell surface receptors: Image of stochastic bonding and rupture processes, in: J. Rabe, H.E. Gaub, P.K. Hansma (Eds.), *Scanning Probe Microscopies and Molecular Materials.* Amsterdam: Kluwer Publishing.

Evans, E., Ritchie, K., 1997. Dynamic strength of molecular adhesion bonds. *Biophys J* 72, 1541–1555.

Fabry, B., Maksym, G.N., Butler, J.P., Glogauer, M., Navajas, D., Fredberg, J.J., 2001. Scaling the microrheology of living cells. *Phys Rev Lett* 87, 148102.

Fabry, B., Maksym, G.N., Butler, J.P., Glogauer, M., Navajas, D., Taback, N.A., Millet, E.J., Fredberg, J.J., 2003. Time scale and other invariants of integrative mechanical behavior in living cells. *Phys Rev E Stat Nonlin Soft Matter Phys* 68, 041914.

Fabry, B., Maksym, G.N., Hubmayr, R.D., Butler, J.P., Fredberg, J.J., 1999. Implications of heterogeneous bead behavior on cell mechanical properties measured with magnetic twisting cytometry. *J Magn Magn Mater* 194, 120–125.

Felsenfeld, D.P., Choquet, D., Sheetz, M.P., 1996. Ligand binding regulates the directed movement of beta1 integrins on fibroblasts. *Nature* 383, 438–440.

Féréol, S., Fodil, R., Labat, B., Galiacy, S., Laurent, V.M., Louis, B., Isabey, D., Planus, E., 2006. Sensitivity of alveolar macrophages to substrate mechanical and adhesive properties. *Cell Motil Cytoskeleton* 63, 321–340.

Féréol, S., Fodil, R., Pelle, G., Louis, B., Isabey, D., 2008. Cell mechanics of alveolar epithelial cells (AECs) and macrophages (AMs). *Respir Physiol Neurobiol* 163, 3–16.

Florin, E.L., Moy, V.T., Gaub, H.E., 1994. Adhesion forces between individual ligand-receptor pairs. *Science* 264, 415–417.

Freedman, A.P., Robinson, S.E., Johnston, R.J., 1980. Non-invasive magnetopneumographic estimation of lung dust loads and distribution in bituminous coal workers. *J Occup Med* 22, 613–618.

Fritz, J., Katopodis, A.G., Kolbinger, F., Anselmetti, D., 1998. Force-mediated kinetics of single P-selectin/ligand complexes observed by atomic force microscopy. *Proc Natl Acad Sci USA* 95, 12283–12288.

Hänngi, P., Talkner, P., Borkovec, M., 1990. Reaction rate Theory: 50 years after Kramers. *Rev Mod Phys* 62, 251–341.

Hinterdorfer, P., Baumgartner, W., Gruber, H.J., Schilcher, K., Schindler, H., 1996. Detection and localization of individual antibody-antigen recognition events by atomic force microscopy. *Proc Natl Acad Sci USA* 93, 3477–3481.

Hu, S., Chen, J., Fabry, B., Numaguchi, Y., Gouldstone, A., Ingber, D.E., Fredberg, J.J., Butler, J.P., Wang, N., 2003. Intracellular stress tomography reveals stress focusing and structural anisotropy in cytoskeleton of living cells. *Am J Physiol Cell Physiol* 285, C1082–C1090.

Ingber, D.E., 1993. Cellular tensegrity: Defining new rules of biological design that govern the cytoskeleton. *J Cell Sci* 104 (Pt 3), 613–627.

Ingber, D.E., 1998. Cellular basis of mechanotransduction. *Biol Bull* 194, 323–325; discussion 325–327.

Isabey, D., Fereol, S., Caluch, A., Fodil, R., Louis, B., Pelle, G., 2013. Force distribution on multiple bonds controls the kinetics of adhesion in stretched cells. *J Biomech* 46, 307–313.

Isabey, D., Pelle, G., Andre Dias, S., Bottier, M., Nguyen, N.M., Filoche, M., Louis, B., 2016. Multiscale evaluation of cellular adhesion alteration and cytoskeleton remodelling by magnetic bead twisting. *Biomech Model Mechanobiol* 15, 947–963. doi:10.1007/s10237-015-0734-5.

Ishijima, A., Harada, Y., Kojima, H., Funatsu, T., Higuchi, H., Yanagida, T., 1994. Single-molecule analysis of the actomyosin motor using nano-manipulation. *Biochem Biophys Res Commun* 199, 1057–1063.

Kramers, H., 1940. Brownian motion in a field of force and the diffusion model of chemical reactions. *Physica* 7, 284–304.

Laurent, V.M., Cañadas, P., Fodil, R., Planus, E., Asnacios, A., Wendling, S., Isabey, D., 2002a. Tensegrity behaviour of cortical and cytosolic cytoskeletal components in twisted living adherent cells. *Acta Biotheor* 50, 331–356.

Laurent, V.M., Fodil, R., Canadas, P., Féréol, S., Louis, B., Planus, E., Isabey, D., 2003. Partitioning of cortical and deep cytoskeleton responses from transient magnetic bead twisting. *Ann Biomed Eng* 31, 1263–1278.

Laurent, V.M., Henon, S., Planus, E., Fodil, R., Balland, M., Isabey, D., Gallet, F., 2002b. Assessment of mechanical properties of adherent living cells by bead micromanipulation: Comparison of magnetic twisting cytometry vs optical tweezers. *J Biomech Eng* 124, 408–421.

Lele, T.P., Sero, J.E., Matthews, B.D., Kumar, S., Xia, S., Montoya-Zavala, M., Polte, T., Overby, D., Wang, N., Ingber, D.E., 2007. Tools to study cell mechanics and mechanotransduction. *Methods Cell Biol* 83, 443–472.

Lenormand, G., Bursac, P., Butler, J.P., Fredberg, J.J., 2007. Out-of-equilibrium dynamics in the cytoskeleton of the living cell. *Phys Rev E Stat Nonlin Soft Matter Phys* 76, 041901.

Li, F., Redick, S.D., Erickson, H.P., Moy, V.T., 2003. Force measurements of the alpha-5beta1 integrin-fibronectin interaction. *Biophys J* 84, 1252–1262.

Maksym, G.N., Fabry, B., Butler, J.P., Navajas, D., Tschumperlin, D.J., Laporte, J.D., Fredberg, J.J., 2000. Mechanical properties of cultured human airway smooth muscle cells from 0.05 to 0.4 Hz. *J Appl Physiol* 89, 1619–1632.

Meyer, C.J., Alenghat, F.J., Rim, P., Fong, J.H., Fabry, B., Ingber, D.E., 2000. Mechanical control of cyclic AMP signalling and gene transcription through integrins. *Nat Cell Biol* 2, 666–668.

Mitrossilis, D., Fouchard, J., Guiroy, A., Desprat, N., Rodriguez, N., Fabry, B., Asnacios, A., 2009. Single-cell response to stiffness exhibits muscle-like behavior. *Proc Natl Acad Sci USA* 106, 18243–18248.

Miyamoto, S., Akiyama, S.K., Yamada, K.M., 1995a. Synergistic roles for receptor occupancy and aggregation in integrin transmembrane function. *Science* 267, 883–885.

Miyamoto, S., Teramoto, H., Coso, O.A., Gutkind, J.S., Burbelo, P.D., Akiyama, S.K., Yamada, K.M., 1995b. Integrin function: Molecular hierarchies of cytoskeletal and signaling molecules. *J Cell Biol* 131, 791–805.

Mizuno, D., Tardin, C., Schmidt, C.F., Mackintosh, F.C., 2007. Nonequilibrium mechanics of active cytoskeletal networks. *Science* 315, 370–373.

Molloy, J.E., Burns, J.E., Kendrick-Jones, J., Tregear, R.T., White, D.C., 1995. Movement and force produced by a single myosin head. *Nature* 378, 209–212.

Munevar, S., Wang, Y., Dembo, M., 2001. Traction force microscopy of migrating normal and H-ras transformed 3T3 fibroblasts. *Biophys J* 80, 1744–1757.

Na, S., Collin, O., Chowdhury, F., Tay, B., Ouyang, M., Wang, Y., Wang, N., 2008. Rapid signal transduction in living cells is a unique feature of mechanotransduction. *Proc Natl Acad Sci USA* 105, 6626–6631.

Ohayon, J., Tracqui, P., 2005. Computation of adherent cell elasticity for critical cell-bead geometry in magnetic twisting experiments. *Ann Biomed Eng* 33, 131–141.

Ohayon, J., Tracqui, P., Fodil, R., Fereol, S., Laurent, V.M., Planus, E., Isabey, D., 2004. Analysis of nonlinear responses of adherent epithelial cells probed by magnetic bead twisting: A finite element model based on a homogenization approach. *J Biomech Eng* 126, 685–698.

Oliver, T., Dembo, M., Jacobson, K., 1999. Separation of propulsive and adhesive traction stresses in locomoting keratocytes. *J Cell Biol* 145, 589–604.

Paszek, M.J., Zahir, N., Johnson, K.R., Lakins, J.N., Rozenberg, G.I., Gefen, A., Reinhart-King, C.A., Margulies, S.S., Dembo, M., Boettiger, D., Hammer, D.A., Weaver, V.M., 2005. Tensional homeostasis and the malignant phenotype. *Cancer Cell* 8, 241–254.

Planus, E., Galiacy, S., Fereol, S., Fodil, R., Laurent, V.M., d'Ortho, M.P., Isabey, D., 2005. Apical rigidity of an epithelial cell monolayer evaluated by magnetic twisting cytometry: ICAM-1 versus integrin linkages to F-actin structure. *Clin Hemorheol Microcirc* 33, 277–291.

Planus, E., Galiacy, S., Matthay, M., Laurent, V., Gavrilovic, J., Murphy, G., Clérici, C., Isabey, D., Lafuma, C., d'Ortho, M.P., 1999. Role of collagenase in mediating in vitro alveolar epithelial wound repair. *J Cell Sci* 112 (Pt 2), 243–252.

Ponton, A., Warlus, S., Griesmar, P., 2002. Rheological study of the sol-gel transition in silica alkoxides. *J Colloid Interface Sci* 249, 209–216.

Potard, U.S., Butler, J.P., Wang, N., 1997. Cytoskeletal mechanics in confluent epithelial cells probed through integrins and E-cadherins. *Am J Physiol* 272, C1654–C1663.

Pourati, J., Maniotis, A., Spiegel, D., Schaffer, J.L., Butler, J.P., Fredberg, J.J., Ingber, D.E., Stamenovic, D., Wang, N., 1998. Is cytoskeletal tension a major determinant of cell deformability in adherent endothelial cells ? *Cell Physiol* 43, C1283–C1289.

Puig-de-Morales, M., Millet, E., Fabry, B., Navajas, D., Wang, N., Butler, J.P., Fredberg, J.J., 2004. Cytoskeletal mechanics in adherent human airway smooth muscle cells: Probe specificity and scaling of protein-protein dynamics. *Am J Physiol Cell Physiol* 287, C643–C654.

Samani, A., Zubovits, J., Plewes, D., 2007. Elastic moduli of normal and pathological human breast tissues: An inversion-technique-based investigation of 169 samples. *Phys Med Biol* 52, 1565–1576.

Schwesinger, F., Ros, R., Strunz, T., Anselmetti, D., Guntherodt, H.J., Honegger, A., Jermutus, L., Tiefenauer, L., Pluckthun, A., 2000. Unbinding forces of single antibody-antigen complexes correlate with their thermal dissociation rates. *Proc Natl Acad Sci USA* 97, 9972–9977.

Stamenovic, D., Coughlin, M.F., 1999. The role of prestress and architecture of the cytoskeleton and deformability of cytoskeletal filaments in mechanics of adherent cells: A quantitative analysis. *J Theor Biol* 201, 63–74.

Stamenovic, D., Coughlin, M.F., 2000. A quantitative model of cellular elasticity based on tensegrity. *J Biomech Eng* 122, 39–43.

Stamenovic, D., Rosenblatt, N., Montoya-Zavala, M., Matthews, B.D., Hu, S., Suki, B., Wang, N., Ingber, D.E., 2007. Rheological behavior of living cells is timescale-dependent. *Biophys J* 93, L39–L41.

Thomas, W.E., Vogel, V., Sokurenko, E., 2008. Biophysics of catch bonds. *Annu Rev Biophys* 37, 399–416.

Tinsley, J.H., De Lanerolle, P., Wilson, E., Ma, W., Yuan, S.Y., 2000. Myosin light chain kinase transference induces myosin light chain activation and endothelial hyperpermeability. *Am J Physiol Cell Physiol* 279, C1285–C1289.

Trepat, X., Grabulosa, M., Puig, F., Maksym, G.N., Navajas, D., Farre, R., 2004. Viscoelasticity of human alveolar epithelial cells subjected to stretch. *Am J Physiol Lung Cell Mol Physiol* 287, L1025–L1034.

Valberg, P.A., 1984. Magnetometry of ingested particles in pulmonary macrophages. *Science* 224, 513–516.

Valberg, P.A., Albertini, D.F., 1985. Cytoplasmic motions, rheology, and structure probed by a novel magnetic particle method. *J Cell Biol* 101, 130–140.

Valberg, P.A., Brain, J.D., 1979. Generation and use of three types of iron-oxide aerosol. *Am Rev Respir Dis* 120, 1013–1024.

Valberg, P.A., Butler, J.P., 1987. Magnetic particle motions within living cells. Physical theory and techniques. *Biophys J* 52, 537–550.

Valberg, P.A., Feldman, H.A., 1987. Magnetic particle motions within living cells. Measurement of cytoplasmic viscosity and motile activity. *Biophys J* 52, 551–561.

Wang, B., Caluch, A., Fodil, R., Fereol, S., Zadigue, P., Pelle, G., Louis, B., Isabey, D., 2012. Force control of endothelium permeability in mechanically stressed pulmonary micro-vascular endothelial cells. *Biomed Mater Eng* 22, 163–170.

Wang, N., 1998. Mechanical interactions among cytoskeletal filaments. *Hypertension* 32, 162–165.

Wang, N., Butler, J.P., Ingber, D.E., 1993. Mechanotransduction across the cell surface and through the cytoskeleton [see comments]. *Science* 260, 1124–1127.

Wang, N., Ingber, D.E., 1994. Control of cytoskeletal mechanics by extracellular matrix, cell shape, and mechanical tension. *Biophys J* 66, 2181–2189.

Wang, N., Ingber, D.E., 1995. Probing transmembrane mechanical coupling and cytomechanics using magnetic twisting cytometry. *Biochem Cell Biol* 73, 327–335.

Wang, N., Tolic-Norrelykke, I.M., Chen, J., Mijailovich, S., Butler, J.P., Fredberg, J.J., Stamenovic, D., 2002. Cell prestress. I. Stiffness and prestress are closely associated in adherent contractile cells. *Am J Physiol Cell Physiol* 282, C606–C616.

Weeks, E.R., Crocker, J.C., Levitt, A.C., Schofield, A., Weitz, D.A., 2000. Three-dimensional direct imaging of structural relaxation near the colloidal glass transition. *Science* 287, 627–631.

Wendling, S., Oddou, C., Isabey, D., 1999. Stiffening response of a cellular tensegrity model. *J Theor Biol* 196, 309–325.

Wendling, S., Planus, E., Laurent, V., Barbe, L., Mary, A., Oddou, C., Isabey, D., 2000. Role of cellular tone and microenvironmental conditions on cytoskeleton stiffness assessed by tensegrity model. *Eur Phys J Appl Phys* 9, 51–62.

Williams, P.M., 2003. Analytical descriptions of dynamic force spectroscopy: Behaviour of multiple connections. *Analytica Chimica Acta* 479, 107–115.

Yoshida, M., Westlin, W.F., Wang, N., Ingber, D.E., Rosenzweig, A., Resnick, N., Gimbrone, M.A., Jr., 1996. Leukocyte adhesion to vascular endothelium induces E-selectin linkage to the actin cytoskeleton. *J Cell Biol* 133, 445–455.

Zaman, M.H., Trapani, L.M., Sieminski, A.L., Mackellar, D., Gong, H., Kamm, R.D., Wells, A., Lauffenburger, D.A., Matsudaira, P., 2006. Migration of tumor cells in 3D matrices is governed by matrix stiffness along with cell-matrix adhesion and proteolysis. *Proc Natl Acad Sci USA* 103, 10889–10894.

Zhu, C., Bao, G., Wang, N., 2000. Cell mechanics: Mechanical response, cell adhesion, and molecular deformation. *Annu Rev Biomed Eng* 2, 189–226.

5

Magnetic Needle Development

David Arnold, Zak Kaufman, and Alexandra Garraud

CONTENTS

5.1 Introduction 94
5.2 Types of Magnetic Needles 94
    5.2.1 Permanent Magnet Needles 95
        5.2.1.1 Description 95
        5.2.1.2 Simulation 96
        5.2.1.3 Advantages 98
        5.2.1.4 Limitations 98
    5.2.2 Electromagnetic Needles 99
        5.2.2.1 Description 99
        5.2.2.2 Simulation 101
        5.2.2.3 Advantages 101
        5.2.2.4 Limitations 102
    5.2.3 Soft-Magnetic Needles Driven by
        Permanent Magnets 102
        5.2.3.1 Description 102
        5.2.3.2 Simulation 103
        5.2.3.3 Advantages 104
        5.2.3.4 Limitations 104
    5.2.4 Summary 105
5.3 Manufacturing Techniques for Magnetic Needles 105
    5.3.1 Bottom-Up Approaches 105
        5.3.1.1 Powder Metallurgical and
        Casting Processes 106
        5.3.1.2 Electrodeposition 106
        5.3.1.3 Physical Vapor Deposition 107
    5.3.2 Top-Down Approaches 107
        5.3.2.1 Laser Machining 107
        5.3.2.2 Electrical Discharge Machining 108
        5.3.2.3 Chemical Etching 108
5.4 Engineering Considerations 109
    5.4.1 Geometry 109
    5.4.2 Material 112

5.5   Applications and Calculations.. 113
      5.5.1   Applying Magnetic Forces... 113
      5.5.2   Magnetic Particle Capture.. 115
5.6   Conclusions... 120
References.. 121

5.1 Introduction

Magnetic needles can be broadly defined as any needle-shaped structures that produce magnetic fields. Generally, the dimensions of a magnetic needle tip would not exceed a few millimeters in size. Their most basic functionality is to apply a magnetic force on magnetic entities. In some instances, this force might have the intent of capturing magnetic nanoparticles. In other cases, the force could be used to manipulate the position of a magnetic structure. Depending on the targeted application, the design of the needle is critical. Choosing the appropriate material and type for the magnetic needle (electromagnet or permanent magnet) can have a large impact in terms of performance and ease of manufacturing.

This chapter begins by describing the three basic types of magnetic needles along with comparative strengths and weaknesses. Finite-element simulations of each type of needle are also provided to visualize the magnetic field patterns produced in each case. Next, some of the common manufacturing techniques used to create small-scale magnetic structures are introduced. Both bottom-up and top-down approaches for fabrication are described. Then, engineering considerations for developing magnetic needles are discussed. These include the optimum geometry of the needle and the material used for manufacturing. Additional simulations of different geometries are included for visualization. Finally, a first-order analytical model of magnetic force is built based on the needle magnetic properties. These calculations are motivated by two of the common applications of magnetic needles, which are applying a magnetic force and capturing magnetic particles.

5.2 Types of Magnetic Needles

Magnetic needles can be characterized based on how their magnetic field is derived. There are three basic options: permanent magnet needles, electromagnetic needles, and soft-magnetic needles driven by permanent magnets. Each of these options has advantages and limitations in terms of performance and manufacturing.

5.2.1 Permanent Magnet Needles

5.2.1.1 Description

Permanent magnets, also known as hard ferromagnets, are materials that retain a residual magnetization and produce a persistent magnetic field without the need for an externally applied magnetic field. This class of magnets includes alnico, ferrite, Sm–Co, and Nd–Fe–B. In contrast, soft magnets are materials that are magnetized in the presence of an external field, but do not retain their magnetization once the external magnetic field is removed. Examples of soft magnets include nickel, iron, cobalt, and many steels, which are attracted by permanent magnets, but do not maintain a magnetic field on their own.

Permanent magnet needles broadly encompass any needle that has a permanent magnet component as the tip. This differentiates this class of needle from soft-magnetic needles driven by permanent magnets, discussed in Section 5.2.3, which have permanent-magnetic components, but not located on the tip. On permanent magnet needles, the magnetic portion is limited to the tip or extended throughout the needle. Figure 5.1 shows some examples of permanent magnet needles. The dark gray color indicates the extent of the permanent magnet portion of the needle, and light gray indicates the remainder of the needle, which can be either a soft-magnetic or nonmagnetic material. The dotted arrows indicate the direction of magnetization. In Figure 5.1a, only a thin layer of the tip is made of hard-magnetic material. Such a configuration can be accomplished by electroplating or sputtering a permanent-magnetic material onto a needle substrate. Figure 5.1b shows a permanent-magnetic tip on a soft or nonmagnetic rod. This configuration could be accomplished by machining or fabricating a hard-magnetic tip and mechanically mounting it onto the rod. The needle in Figure 5.1c is entirely made totally of a hard-magnetic material.

An example of a permanent-magnetic tip on a soft-magnetic tip is shown in Figure 5.2. A bulk Nd–Fe–B magnet has been laser-micromachined into a conical shape. Figure 5.2a shows an image of the conical needle and Figure 5.2b shows a picture of the magnetic needle tip attached to a soft-magnetic steel rod

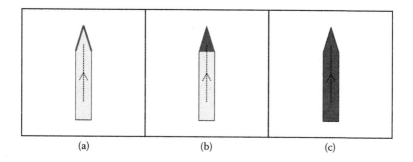

| (a) | (b) | (c) |

FIGURE 5.1

(a–c) Examples of permanent magnet needles. The dark gray color indicates hard, permanent-magnetic material, and light gray indicates soft-magnetic or nonmagnetic material.

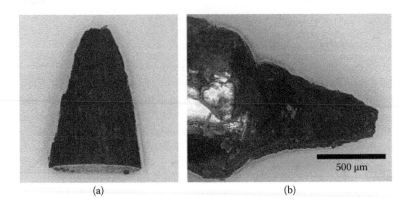

(a)   (b)

FIGURE 5.2
(a) Permanent magnet tip and (b) mechanically attached onto a soft-magnetic rod.

using an adhesive. In this case, the assembly process is as follows: An adhesive was applied to the tip of the rod. Then, the tip of the rod was brought near the base of the permanent magnet cone. Due to the natural magnetic attraction between the hard-magnetic and the soft-magnetic rod, the tip remained in place while the adhesive dried to provide a more permanent bond. This self-assembly aids the creation of the needle structure.

5.2.1.2 Simulation

Finite-element simulations were performed using COMSOL Multiphysics 4.3 to determine the magnetic stray field distribution from a permanent magnet needle. The study is performed assuming a needle of the type presented in Figure 5.1b. In this case, the tip of the needle is a permanent-magnetic material with a 1-T remanent flux density and a relative permeability of 1. These values approximate the properties of bulk Nd–Fe–B. The rod of the needle is simulated as soft-magnetic iron, considering the magnetic hysteresis curve of this material.

In Figure 5.3, color shades indicate magnitude, and arrows indicate direction. For this permanent-magnetic needle, the maximum magnetic flux density is 1.12 T and occurs in the tip of the needle, as expected due to geometry (see Section 5.4). Likewise, the strongest gradients of the flux density occur near the tip. The gradient of flux density at the interface of the tip and air is valued at 1.4×10^5 T/m.

In this simulation, and all others in this chapter, finite-element analysis (FEA) has been used, using COMSOL Multiphysics. In FEA, the geometry of the simulation is broken down into discrete elements of finite size, called the mesh. Each element in the mesh has characteristic equations related to the overall simulation. When the simulation is run, the software attempts to simultaneously solve all the mesh equations while reducing the overall error. This results in the simulation being a good approximation of what happens; however, certain artifacts of the mesh can sometimes be seen in

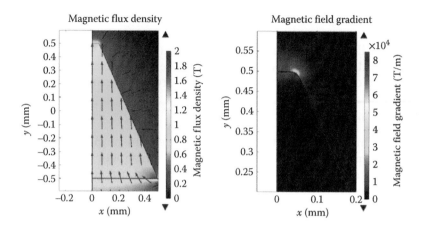

FIGURE 5.3

Simulations of magnetic flux density and magnetic field gradient for a permanent-magnet needle of the type in Figure 5.1b.

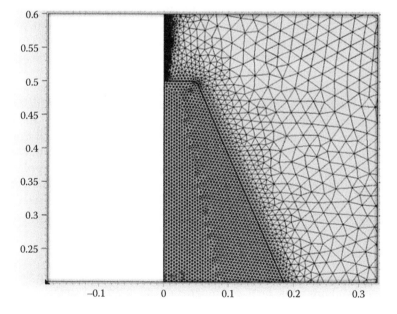

FIGURE 5.4

The mesh used in the simulations of Figure 5.3.

the simulation results. For instance, at boundaries, such as that between the needle tip and air, the results have a significant margin of error.

The sizes of the mesh elements play a role in the extent of the simulation errors. Therefore, the mesh sizes are generally made smaller near boundaries to reduce these errors. Figure 5.4 shows the mesh used to obtain

Figure 5.3. Note that all the simulations in this chapter have been done using similar mesh.

Additionally, it should be noted that the sharp corners have been modeled as round corners in these simulations to prevent mathematical anomaly. These sharp corners could create a source of error in the simulations.

5.2.1.3 Advantages

One of the most attractive features of using permanent magnet needles is that they require no external field to produce a magnetic field. Since permanent magnets sustain their own magnetic field, no electric current is required to provide the field, as is the case for electromagnets. This offers many advantages. For one, portability is not an issue since no external power supply is needed to power the magnet. This makes the needle easier to move and control without the need for wires tethering it down.

Electromagnets scale unfavorably since the produced magnetic field is the result of an electric current which dissipates more heat as the wires scale to smaller sizes.[1] Heat created by electromagnetic needles can be exceedingly high after only seconds of operation.[2] On the contrary, permanent magnets can be scaled down to small sizes and retain a proportional magnetic field, and no heat is generated by the permanent magnet. For applications inside of living organisms, even a small amount of heat can harm or kill cells. In such a situation, permanent magnet needles have a clear advantage.

Finally, for permanent magnet needles, the source of the magnetic field is directly at the tip of the needle. This is in contrast with soft-magnetic needles and electromagnetic needles where the field is often generated on the rod of the needle and then guided through the tip. Producing the field at the tip is advantageous because it allows the field to be localized, if desired, rather than spanning throughout the entire needle. This can be beneficial for applications requiring locational precision of the magnetic field. Additionally, producing the field at the tip avoids the possibility of losing some of the field strength as in the case where the field is produced elsewhere and guided through the tip.

5.2.1.4 Limitations

Despite the many advantages of permanent magnet needles, there are some fundamental limitations. First, the magnetic field is limited to the strength of the magnet. Depending on the magnetic material, when no external field is present, the magnetic flux density cannot exceed the remanent magnetic flux density that can be maintained by the magnet. Most permanent magnet materials have remanence values about ~1 T, with the strongest magnets reaching values of 1.4 T.[3] These values, however, correspond to the bulk material. Unfortunately, when the bulk material is machined down to the size of a needle, significant magnetic damage may potentially occur.[4] Additionally, creating permanent magnet layers through bottom-up approaches, such as electroplating, generally does not yield magnetic properties as strong as the bulk material.[3] Therefore, for needle applications, the field strength produced by permanent magnets can be limited, depending on the manufacturing process. Additionally, to initially magnetize the thin-film permanent magnets, a strong magnetic field must be applied, generally from a pulse magnetizer, to magnetize the magnet.

Another issue with permanent-magnetic needles arises due to the inability to turn the magnetic field off. Unlike electromagnets and soft-magnetic needles driven by permanent magnets, which can be mostly demagnetized by removing the source of magnetization, permanent magnets retain their magnetic field. Permanent magnets are not well-suited for applications that require both capture and release of magnetic particles. One potential way to circumvent this problem is to demagnetize the permanent magnet through other means such as heating it beyond the material Curie point or applying a magnetic field produced by an alternating current; however, these methods would generally occur outside of the experimental environment, so they would only be applicable for subsequent removal of magnetic particles, not for precise movement of particles within the environment.

One final consideration when working with permanent magnets is that it can be difficult to manufacture small geometries. Rare-earth magnets, which are the most powerful permanent magnets, are very brittle and delicate to corrosion.[3] These magnets are generally coated with nickel, to protect them from oxidation. Traditional mechanically based manufacturing techniques will not work for small, permanent magnets without damaging them. Laser machining is a potential alternative for producing small permanent magnets, but has its limitations as well. This is discussed in more detail in Section 5.3.

5.2.2 Electromagnetic Needles

5.2.2.1 Description

Electromagnets can be described as magnets in which the magnetic field is produced by a wire carrying an electric current. The wire is usually wound around a soft-magnetic core such as iron, to amplify and concentrate the magnetic flux. A diagram of an electromagnetic needle is shown in Figure 5.5. In this diagram, there is a coil of wire wound around the shaft of the needle, near the tip. The outer boxes show the direction of the electric current in the wires, where X indicates into the page and circles indicate out of the page. The dotted line indicates the resulting on-axis, direction of the magnetic field.

The field produced by an electromagnet is proportional to the current in the wire until the magnetic core is saturated. Before saturation, increases in the current produce a magnetic field which is amplified by the soft-magnetic core. Once the core is saturated, further increases in current will only slightly increase the magnetic field, since the core does not amplify

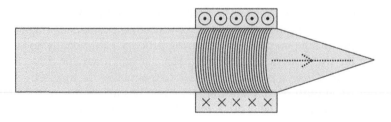

FIGURE 5.5
Electromagnetic needle.

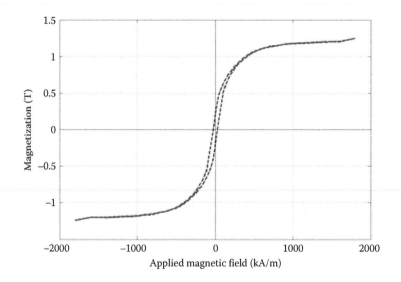

FIGURE 5.6
Example of the magnetization curve of an electromagnetic needle.

additional increase in the field. The magnetization curve (also known as hysteresis curve or M-H curve) is a good way to visualize the relationship between applied current and the resulting magnetization.

Figure 5.6 shows an example of a magnetization curve for a magnetic material. The H-field, or magnetizing field, is the magnetic field created by the electric current, and its value is independent of the material the field is in. The M-field is the magnetization, expressed in T, that arises in a material for a given applied magnetizing H-field. This M-field arises because of the magnetic domains within the magnet aligning to match the direction of the applied magnetizing field. As seen in the curve, after the M-field reaches about 1 T, further increases in the H-field do not significantly increase the magnetization. This is a result of the magnetic saturation phenomenon.

5.2.2.2 Simulation

A simulation of an electromagnetic needle is provided in Figure 5.7. In this simulation, the needle is simulated as soft-magnetic iron surrounded by current-carrying loops of wire. The loop of wire is assumed to have 1000 turns of wire with a conductivity of 6×10^7 S/m and a cross-sectional area of 2.5×10^{-9} m^2. The wire's current is simulated as 0.2 A. These values model an electromagnetic needle found in the literature.[2]

In Figure 5.7, color shades indicate magnitude, and arrows indicate direction. The same color scale as Figure 5.3 is used for easy comparison. For this electromagnetic needle, the maximum magnetic flux density is 1.70 T and occurs at the tip of the needle, as expected due to geometry (see Section 5.4). Likewise, the strongest gradients of the flux density occur near the tip. The gradient of flux density at the interface of the tip and air is valued at 1.23×10^5 T/m.

5.2.2.3 Advantages

Electromagnetic needles provide some advantages over the two other types of needles. First, the magnetic field from an electromagnetic needle can be mostly turned off by switching off the current. If a soft-magnetic core is used, a small remanent field would remain even after the current is removed. The ability to turn the field on and off allows the electromagnetic needles to perform both capture and release of magnetic particles. This can be used to relocate particles within an environment without having to remove them.

Another advantage of electromagnets is that the magnetic field can be controlled in magnitude by varying the current. This is demonstrated in Figure 5.6, since the H-field is directly related to the current. The relationship

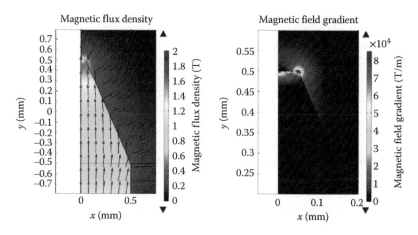

FIGURE 5.7
Simulations of magnetic flux density and magnetic field gradient for an electromagnetic needle of the type in Figure 5.5.

between the M-field and the current is no longer linear at large currents, but it is close to linear until saturation is reached. This allows for electromagnetic needles to perform experiments where a precise magnetic field or force needs to be applied.

Finally, electromagnetic needles are generally able to produce the highest magnetic fields of the three needle options. Indeed, the field produced by a permanent magnet is limited to the remanence of the material, as discussed earlier. For electromagnets, however, the field can theoretically be increased up to the saturation magnetization of the core material. As seen in the simulations (Figure 5.7), the electromagnetic needle produces both the highest flux density and flux density gradient of the three needle types. If heat and power are not an issue, electromagnets can produce the strongest fields.

5.2.2.4 Limitations

One major limitation of electromagnetic needles, especially for biological applications, is that they can produce a high heat dissipation. The literature shows that an electromagnetic needle can experience a 14°C temperature rise within 8 seconds of operation.[2] This measure was for an electromagnetic needle containing 500 turns, a 0.7-A current, and 50-μm diameter wire. This temperature increase is a major issue for *in vivo* applications, where small increase in temperature could lead to damage in tissue. For applications outside of a living host, the heat may be a smaller concern. One potential solution to mitigate the heat issue is to build a temperature regulation system around the needle, as done in the literature.[2] Another workaround is to use the needle only in short pulses, or to use it at low currents, which may produce a strong enough field for some applications.

Another potential limitation of electromagnets is that they require a power source to provide the current. This should not be limiting for most applications, but it could put constraints on portability.

Lastly, the magnetic fields produced by electromagnets are less localized than fields that can be produced with permanent-magnet needle tips. In the case of the electromagnet, the magnetic field is being produced by a coil of wire, located further down the needle, away from the tip. This means the field for electromagnets is often spread throughout the needle, covering a larger volume.

5.2.3 Soft-Magnetic Needles Driven by Permanent Magnets

5.2.3.1 Description

The final type of magnetic needle is a soft-magnetic needle driven by a permanent magnet. In this type of needle, the tip of the needle is a soft magnet instead of a permanent magnet, but at least another portion of the needle is made of a permanent magnet. This differentiates this type of needle from

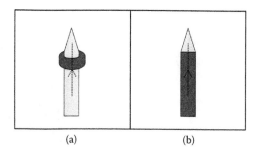

(a)   (b)

FIGURE 5.8
(a) and (b) Examples of soft-magnetic needles driven by permanent magnets. The dark gray indicates hard, permanent-magnetic material, and gray indicates soft-magnetic material.

permanent magnet needles, described in the previous section, which have a permanent magnet tip.

Two potential configurations for soft-magnetic needles driven by permanent magnets are shown in Figure 5.8. In Figure 5.8a, the entire needle is made of a soft-magnetic material that does not produce any field by itself. The dark gray ring around the needle is a permanent magnet that creates a magnetic field that is amplified and guided through the soft-magnetic needle. This configuration is useful because the magnetic field can be turned off by removing the permanent magnet ring. Figure 5.8b presents a needle that is comprised of a permanent magnet shaft with a soft-magnetic tip. Such a needle could be produced by mechanically attaching the soft-magnetic tip or by manufacturing the tip on top of the shaft through electroplating or other bottom-up approaches. The dotted lines in the figures indicate the direction of magnetization.

5.2.3.2 Simulation

A simulation of a soft-magnetic needle driven by a permanent magnet is provided in Figure 5.9. This is a needle of the type in Figure 5.8b. In this simulation, the base of the needle is a permanent-magnetic material with a remanent flux density of 1 T and a relative permeability of 1. These values approximate the properties of bulk Nd–Fe–B. The tip of the needle is simulated as iron.

In Figure 5.9, color shades indicate magnitude, and arrows indicate direction. The same color scale as for the previous two simulations are used for easy comparison. For this soft-magnetic needle, the maximum magnetic flux density is 1.40 T and occurs at the tip of the needle, as expected due to geometry (see Section 5.4). Likewise, the strongest gradients of the flux density occur near the tip. The gradient of flux density at the interface of the tip and air is valued at 1.5×10^4 T/m.

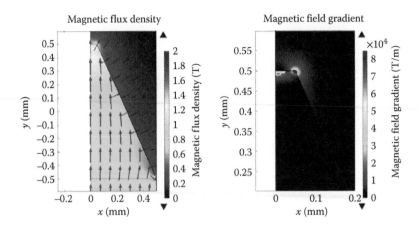

FIGURE 5.9
Simulations of magnetic flux density and magnetic field gradient for a soft-magnetic needle of the type in Figure 5.8b.

5.2.3.3 Advantages

The most practical advantage of soft-magnetic needles driven by permanent magnets is that they can be easier to manufacture than permanent magnet needles. Other than this, they share many of the same characteristic advantages of permanent magnet needles. Namely, they do not require an electric current to produce their magnetic field. This allows for the same portability and absence of heat generation provided by permanent magnet needles.

An extra benefit of using soft-magnetic needles driven by permanent magnets is that they can be turned off or reduced in strength if they are the type of configuration shown in Figure 5.8a. When the magnetic ring is removed from the needle, the magnetic field mostly vanishes. Permanent magnet needles do not share this advantage, as they are always on.

Manufacturing soft-magnetic needles is often easier than manufacturing permanent magnet needles. Rare-earth permanent-magnetic materials are brittle and can be difficult to manufacture into small geometries, particularly in the case of a needle tip. Since the soft-magnetic needle driven by a permanent magnet does not have a permanent-magnetic tip, it avoids this manufacturing difficulty.

5.2.3.4 Limitations

Some of the same limitations of permanent magnet needles apply to soft-magnetic needles driven by permanent magnets. The strength of the magnetic field is limited by the magnetic materials used for the permanent magnet and the soft-magnetic tip. Additionally, manufacturing small, permanent magnets can still be difficult, even if the tip is not a permanent magnet.

TABLE 5.1

Summary of Needle Types

	Permanent Magnet	Electromagnet	Soft Magnet Driven by PM
Manufacturing	Difficult	Easy	Moderate
Field Strength	Moderate	Strong	Moderate/weak
Gradient	Moderate	Strong	Moderate/weak
Heat Produced	No	Yes	No
External Power	No	Yes	No
Variable Field	No	Yes	Sometimes
Field Removable	No	Yes	Sometimes
Scales Favorably	Yes	No	Yes

One additional limitation not shared by permanent magnet needles is the fact that the source of the field is not directly produced at the tip of the needle. Rather, it is produced further down the shaft and guided through the tip. This generally leads to a weaker gradient at the tip interface, as observed in the simulation results.

5.2.4 Summary

In summary, the three needle types have various advantages and limitations. Depending on the application, one needle type may be more desirable than another. The characteristics of the three needle types are summarized in Table 5.1.

5.3 Manufacturing Techniques for Magnetic Needles

This section provides a brief overview of potential manufacturing techniques for magnetic needles for readers who are unfamiliar with microfabrication techniques. A more thorough treatment of methods for manufacturing small-scale magnets can be found in Refs. (3,5).

Most of the following manufacturing processes can be used to create either permanent magnets or soft magnets, depending on the material used. Techniques for manufacturing magnets can be broken into two categories: bottom-up approaches and top-down approaches.

5.3.1 Bottom-Up Approaches

Bottom-up manufacturing brings small parts together into a larger product. In the case of magnetic needles, this involves bringing magnetic powders or magnetic elements together to form a larger magnet. Three common types of bottom-up approaches include powder metallurgical and casting processes, electrodeposition, and physical vapor deposition (PVD).

5.3.1.1 Powder Metallurgical and Casting Processes

Most bulk (large) permanent magnets are produced using powder metallurgical or casting processes. These processes begin by melting the raw material elements into a molten mixture with stoichiometric ratio for the desired magnetic material. For casting processes, the mixture is poured into a mold and cooled until it solidifies. For powder processes, the cooled mixture is grinded into a powder. This powder is then mixed with an organic binder to eventually create bonded magnets. This step is required because many magnetic materials are brittle on their own. Therefore, the binder allows them to be extruded or molded into various geometries without breaking.[3]

These powder and casting processes are widespread in industry to manufacture large-scale magnets; however, the processes can be extended to work on small-scale magnets, such as those used in magnetic needles. For example, the powder technique has been successfully scaled for submillimeter-sized magnets, using wax as the bonding material.[6]

5.3.1.2 Electrodeposition

Electrodeposition is a bottom-up approach that involves plating metallic materials onto a conductive substrate. An aqueous bath is prepared containing ions of the material(s) to be electroplated. Then, an anode and cathode are inserted into the bath. The anode is connected to the negative terminal of a power supply, and the cathode is connected to the positive terminal. This results in electrons flowing through the bath from the anode to the cathode. When the electrons reach the cathode, they oxidize the positively charged ions in the bath to give them a neutral charge and deposit them onto the cathode.[7]

In this scheme, the substrate to be deposited on is located at the cathode electrode, and it requires the presence of a conductive layer. The anode can either be consumable or nonconsumable. In the consumable case, the anode is dissolved throughout the process and provides ions to the bath. In the nonconsumable case, ions must be available in the bath, and replenished when most of them have been plated onto the cathode.

The most common magnetic materials that are electrodeposited are metal-alloy magnets. Some examples of metal-alloy magnets that have been electroplated include Co–Pt, Fe–Pt, and various alloys containing Ni–Co along with a nonmetallic element. Unfortunately, these metal-alloy magnets tend to have weaker magnetic properties than rare-earth magnets.[3] It limits the utility of electrodeposited magnetic materials in the application of magnetic needles. It should be noted that although rare-earth magnets cannot be electroplated in aqueous baths, it is possible to electroplate these materials in molten salt baths.

Another limitation of electrodeposition is that the thickness of the deposited films is generally limited to a few micrometers because of mechanical stresses that develop during the deposition process. Additionally, as the film grows, grains will begin to form, which decreases the uniformity of the magnet.[3]

5.3.1.3 Physical Vapor Deposition

PVD processes include sputtering, evaporation, and pulsed-laser deposition. In these processes, the material to be deposited on the substrate comes from a target. Then, it is condensed back to its solid form on the substrate after traveling in a vacuum chamber. This is a purely physical reaction.[7]

One great advantage of PVD processes is that they can be used to deposit rare-earth magnetic materials. Numerous groups have been able to use sputtering to produce Sm–Co and Nd–Fe–B rare-earth magnets with properties approaching those of the bulk materials.[3] One limitation of these sputtered rare-earth magnets is that it can be difficult to shape them; however, wet-etching has been shown as a reasonable method for shaping these magnets.[8]

As electrodeposition, PVD suffers from the inability to produce large thickness of deposited material due to mechanical stresses that develop in the material as the thickness increases.

5.3.2 Top-Down Approaches

As described in the previous section, electrodeposition and PVD approaches are limited in that they cannot produce magnets over tens of micron in thickness. Depending on the application, this may present a problem, particularly for magnetic needles, which may require a significant thickness in some applications.

An alternative to these bottom-up approaches are top-down approaches where a bulk material is reduced to the desired size through a subtractive process. Three such top-down approaches include laser machining, electrical discharge machining (EDM), and chemical etching.

5.3.2.1 Laser Machining

One particularly promising method for producing magnetic needles is laser machining. In this process, a high-energy laser is used to ablate material from a bulk substrate to produce the desired shape. This process has the potential to produce complex shapes of magnetic needle, allowing for control of the magnetic field pattern. Additionally, this process allows for a rapid method of producing needles. The magnetic needle tip in Figure 5.2

has been fabricated by laser machining a conical structure out of a bulk substrate of Nd–Fe–B.

Laser machining can be compared to mechanical machining in that they both allow for complex shapes and relatively rapid production; however, in the case of rare-earth magnetic materials, laser machining provides an advantage. Since rare-earth magnets are very brittle, mechanically machining them can cause them to fracture. Laser machining avoids this issue.

One limitation of laser machining is that the continuous input of energy to the sample causes the material to reach high temperatures if not regulated. For permanent magnets, high temperatures can result in demagnetization and sometimes permanent-magnetic damage due to oxidation and internal changes. This ultimately means that the laser-machined magnet would have weaker magnetic properties than the bulk material. This heat effect can be potentially mitigated by lowering the laser power, by introducing cooling cycles at intervals throughout the laser machining process, or by flowing a cool gas over the surface to dissipate heat.

5.3.2.2 Electrical Discharge Machining

EDM is a manufacturing process where a bulk material is manufactured into a desired shape by using a series of electrical discharges. In this process, a large electric field is established between an electrode on the tool and the sample to be machined. Generally, a liquid dielectric is placed between the electrode and the sample. As the tool electrode is brought near the sample, the electric field becomes strong enough to break down the dielectric material, resulting in a current flow. As this breakdown current flows, some of the material is removed from both the tool electrode and the sample, near the position of the tool electrode. This method of machining is similar to laser machining because it can be used to machine a bulk magnet into complex shapes. The heat produced in this manufacturing technique poses similar problems as in the case of laser machining, but similar methods to control the heat can be used.

5.3.2.3 Chemical Etching

Chemical etching is a third option for top-down manufacturing of magnetic needles. In this process, a chemical etchant, typically an acid, is used to dissolve portions of a bulk substrate to produce the desired shape.[7] This method often requires a preliminary patterning step, before immersion in the etchant so that only selective portions of the material would be removed.

In the literature, both Nd–Fe–B and Sm–Co magnetic films have been etched using acidic baths containing $(NH_4)_2S_2O_8$, H_2O, and H_2SO_4. This etching recipe achieved removal rates of 1.25 μm/min and relatively vertical side walls.[5]

5.4 Engineering Considerations

When designing magnetic needles, there are various engineering consider-ations that should be considered depending on the required application of the needle. The two considerations discussed in this section are the geom-etry and the material of the needle.

5.4.1 Geometry

The geometry of a magnetic needle has a large impact on the magnetic behav-ior. The magnetic flux produced by a magnetic needle must be conserved. As discussed previously, the flux can either be produced by a permanent magnet or an electromagnet. Either way, the flux remains relatively constant throughout the needle. The magnetic flux density, however, can change by changing the cross-sectional area of the needle:

$$\mathbf{B} = \frac{\Phi}{A} \tag{5.1}$$

Equation 5.1 shows that the magnetic flux density, **B**, can vary with the cross-sectional area, A, even though the flux, Φ, remains constant. This means that a tapered-shaped needle tip can concentrate the flux into a small area, thereby increasing the flux density, measured in T. This concept is illus-trated in Figure 5.10.

Consequently, the narrowing of the needle tip allows for producing very high magnetic flux densities and flux density gradients. It will be shown in Section 5.5 that the magnetic force that a magnetic needle can apply is directly related to the gradient of the flux density. Therefore, a needle with the geometry of Figure 5.10 should produce a strong force on a magnetic particle near the tip.

Although tapered needles produce high flux density gradients, their reach is limited to a smaller volume. Matthews et al.[2] have shown that increasing

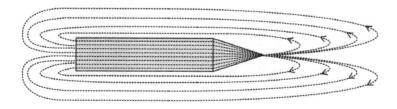

FIGURE 5.10
Magnetic flux lines are concentrated by the geometry of the needle tip.

the tapering of a magnetic needle tip causes a shaper cutoff in the force versus distance curve. As the needle tip is extended, the distance over which the force extends decreases.

Figures 5.11 through 5.13 show simulations for three different electromagnetic needle geometries. Each one has a different degree of tapering. In each case, the needle is simulated with the same properties as the electromagnetic needle simulated in Section 5.2.2. Each tip has a base radius of 0.5 mm and a tip radius of 0.05 mm. The first needle tip has a height of 0.5 mm, the second has a height of 1.0 mm, and the third has a height of 2.0 mm. The magnetic field in the soft magnet is notably increased at the tip location which increases the magnetic field gradient around it. Furthermore, the magnetic field gradient is large over a larger region when the taper becomes longer and longer.

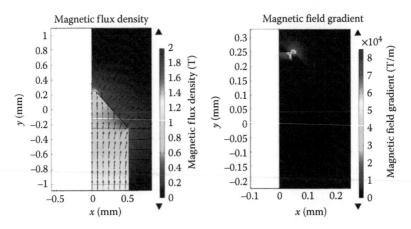

FIGURE 5.11
Simulations of magnetic flux density and its gradient for needle with short taper.

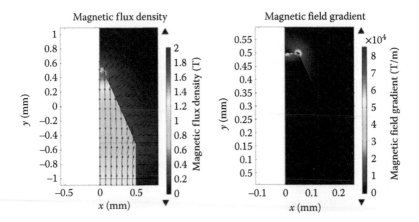

FIGURE 5.12
Simulations of magnetic flux density and its gradient for needle with medium taper.

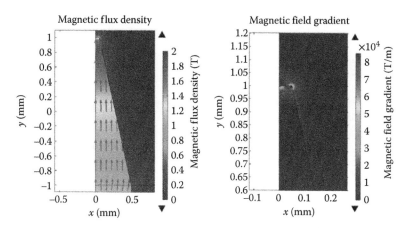

FIGURE 5.13
Simulations of magnetic flux density and its gradient for needle with long taper.

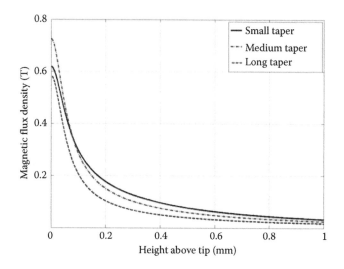

FIGURE 5.14
Magnetic flux density above tip for three different amounts of tapering.

Figures 5.14 and 5.15 summarize the results from the three simulations by plotting the flux density and the magnetic field gradient as a function of the distance from the tip of the needle, on-axis. As seen from these simulations, as the tip becomes longer, the magnetic flux density extending from the tip of the needle decreases. Additionally, the gradient of the flux has a sharper roll-off for the longer needles, meaning that they would apply a strong force over a shorter distance. This reduced reach of the force can be useful for actuating a small set of particles within a larger group. This trend matches

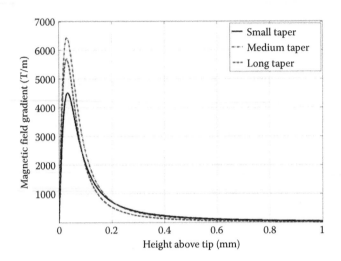

FIGURE 5.15
Magnetic flux density gradient above tip for three different amounts of tapering.

experimental results found in the literature,[2] but the experimental results were even more pronounced than those from these simulations.

5.4.2 Material

The magnetic material chosen for creating small magnetic structures can have a significant impact on the magnetic performance and the usability. Ferromagnetic materials can be divided into two general categories: permanent magnets and soft magnets. Permanent magnets retain their magnetic field without the presence of an external field. Soft magnets can be magnetized in the presence of an external field, but they lose most of their magnetization when the external field is removed.

Common permanent magnet materials can be classified into three groups: ferrites, transition-metal alloys, and rare-earth alloys. These are listed in increasing order of energy density.[3]

Ferrites are commonly used as general-purpose magnets. They provide excellent corrosion resistance and low cost of manufacturing. However, they provide relatively poor magnetic properties. These magnets typically have remanence values less than 0.5 T. Some common ferrites include $BaO \cdot 6(Fe_2O_3)$, $SrO \cdot 6(Fe_2O_3)$, and $PbO \cdot 6(Fe_2O_3)$. Although these materials are technically ferrimagnetic, they behave similarly to ferromagnets on a large scale.[3]

Transition-metal alloys pair ferromagnetic transition-metal elements such as Fe, Co, and Ni with nonferromagnetic metals including Al, Pt, and Cr. Some examples include Al–Ni–Co (Alnico) magnets, Co–Pt, and Fe–Pt. These classes of magnets typically have high remanence values in the order of 1 T and can withstand high temperatures, but they suffer from low to mid-coercivity, meaning they can potentially be demagnetized.[3]

The magnets with the strongest ferromagnetic behavior are rare-earth alloys. The most commonly available rare-earth magnets are Nd–Fe–B and Sm–Co. These magnets can have remanence values above 1 T and high coercivity. Their primary disadvantage is that they are prone to oxidation and cannot withstand high temperatures.[3] To prevent oxidation, these magnets generally must be coated with a protective layer, such as nickel. Another limitation of these materials is that they are very brittle.

Considering these three classes of permanent-magnetic materials, some conclusions can be drawn based on the required application of a magnetic needle. When possible, rare-earth magnets should generally be chosen to maximize the magnetic performance of the needle. If the needle must withstand high temperatures larger than 125°C, transition-metal alloys offer a good compromise between magnetic performance and high operating temperature. Transition-metal alloy magnets and ferrites might also be considered in cases where the brittleness of rare-earth magnets poses a manufacturing problem.

5.5 Applications and Calculations

In this section, some applications of magnetic needles are discussed and calculations are derived as they apply to them. The two primary applications of magnetic needles are applying magnetic forces and capturing magnetic particles. These calculations will provide a theoretical framework for understanding the utility of magnetic needles.

5.5.1 Applying Magnetic Forces

The primary application of a magnetic needle is to apply magnetic forces. This type of force can be used for a variety of applications, including guiding magnetic particles, measuring intracellular forces, and for magnetic gene transfection. In these cases, knowledge of the magnetic force being applied by the magnetic needle is of importance. The magnetic particles that would be manipulated are nano- to microscale magnetic particles. In consequence, in a first-order approximation, we can assume that the magnetic force is being applied to a magnetic dipole.

The equation for the magnetic force on a magnetic dipole can take two forms, depending on the dipole model used in the derivation.[9] The electric current loop model for a magnetic dipole yields

$$F_m = \nabla (m \cdot B) \tag{5.2}$$

Alternatively, the separated magnetic charge model for a magnetic dipole yields

$$F_m = (m \cdot \nabla) B \qquad (5.3)$$

In both equations, B is the applied magnetic field from the needle, measured in T, and m is the magnetic moment of the dipole the force is being applied to, in A m². In practice, under some assumptions (magnetic dipole, no external current), the two equations return similar results,[9,10] so the more common (Equation 5.2) will be used in this section.

If the magnetic force acts on a permanent magnet, then Equation 5.2 is the result for the magnetic force since m and B are independent in this case. However, when the magnetic force is applied on a soft-magnetic particle, m will become magnetized in the same direction as B. Further, when the dipole is close in distance to the magnetic needle, it will achieve the magnetic saturation state. In some applications of magnetic needles, where a force is being applied on superparamagnetic nanoparticles, these conditions are generally met. These conditions allow the following simplifications to be made.

The magnetic moment of the magnet m, in A m², is defined as

$$m = M \cdot \mathrm{vol} \qquad (5.4)$$

In this equation, vol is the magnetic volume of the material, in m³, and M is the magnetization of the material, in units of A m⁻¹.

Further, when the magnetic particle is fully saturated, its magnetization is proportional to B_s, the saturation flux density of the magnetic particle, in T, as

$$M = \frac{1}{\mu_0} B_s \qquad (5.5)$$

Note that this equation assumes the particle is magnetically saturated. In the case of nonsaturation, the magnetization is written as

$$M = \frac{1}{\mu_0} \chi_m B \qquad (5.6)$$

In this case, χ_m is the volume magnetic susceptibility, which is a measure of how much a magnetic material amplifies an applied magnetic field. This derivation will ignore the linear behavior part and only assume saturation. By substituting Equation 5.5 into Equation 5.4 and substituting the result into Equation 5.2, the force is found as

$$F_m = \nabla \left(\frac{\mathrm{vol}}{\mu_0} B_s \cdot B \right) \qquad (5.7)$$

Finally, the B_s vector must be in the same direction as the B vector in the case of a soft-magnetic particle, because the external field magnetizes the magnet in the same direction. Since the vectors are in the same direction, the dot product becomes a simple product of vector magnitudes as in Equation 5.8. Then, all the constants are brought in front of the gradient for the result of Equation 5.9.

$$F_m = \nabla\left(\frac{\text{vol}}{\mu_0}B_sB\right) \tag{5.8}$$

$$F_m = \frac{\text{vol}}{\mu_0}B_s\nabla(B) \tag{5.9}$$

This result shows that the force on a soft-magnetic particle from a magnetic needle is proportional to the volume of the particle, the saturation magnetic flux density of the particle, and the gradient of the magnitude of the applied magnetic field from the needle. This result indicates that a higher magnetic force can be applied to a magnetic particle by not only increasing the strength of the magnetic field, but by increasing the gradient. The gradient can be controlled by the geometry of the magnetic needle, as discussed in Section 5.4.

The results derived in this section are useful for applications such as applying magnetic forces on magnetic particles or measuring intracellular forces by attaching magnetic nanoparticles to structures in a cell and determining the magnetic force required to balance the intracellular force.

5.5.2 Magnetic Particle Capture

Another application of magnetic needles is the capture of magnetic particles. Working with the results from the previous section, some useful derivations can be made in relation to particle capture. This section provides a first-order approximation for explaining the main parameters. More details in the derivation can be found in Refs. (11,12).

When a superparamagnetic nanoparticle in a fluid is subjected to an external magnetic field gradient, it will experience a magnetic force as given by Equation 5.9. Once the particle begins to move, it will also experience a drag force in the opposite direction of its motion, given as

In this equation, η is the viscosity of the fluid, in Pa s, D is the hydrodynamic diameter of the particle, in m, and v is the particle velocity, in m s^{-1}.

$$F_d = -3\pi\eta Dv \tag{5.10}$$

At any given distance from the magnetic needle, the particle will be traveling at a velocity such that the drag force and the magnetic force are equal in magnitude but opposite in direction. By equating the magnetic force and the drag force, an expression for the velocity can be determined. If the magnitudes of Equations 5.9 and 5.10 are equated, the velocity is found to be as

$$v = \frac{\text{vol} \cdot B_s}{3\pi\mu_0\eta D} \nabla(B) \tag{5.11}$$

This result can be simplified by only considering particles traveling in a straight line coming toward the tip of the needle on-axis, as shown in Figure 5.16.

If this is considered the z-dimension, then the single-dimensional velocity can be described as

$$v_z(z) = \frac{\text{vol} \cdot B_s}{3\pi\mu_0\eta D} \frac{dB(z)}{dz} \tag{5.12}$$

This result gives an equation for the velocity experienced by a particle on-axis with a magnetic needle at a given distance, z. With this result, the time of flight, T, for a magnetic particle can be determined. The time of flight is defined as the amount of time it takes for the particle to travel from an initial distance, z, until it reaches the magnetic needle tip.

By rearranging the equation for the definition of velocity, a differential equation can be created for the time-of-flight determination:

$$v_z(z) = \frac{dz}{dt}$$
$$dt = \frac{dz}{v_z(z)} \tag{5.13}$$

Integrating both sides, the time of flight for a general velocity function is found as

$$T(z) = \int_z^0 \frac{1}{v_z(z)} dz = \int_0^z \frac{-1}{v_z(z)} dz$$

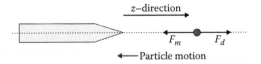

FIGURE 5.16
Coordinate system for particle approaching needle on-axis.

Next, Equation 5.12 is substituted for the velocity to obtain a general time-of-flight formula:

$$T(z) = \int_0^z \frac{-1}{\dfrac{\text{vol} \cdot B_s}{3\pi\mu_0\eta D} \dfrac{dB(z)}{dz}} dz$$

(5.14)

$$T(z) = \frac{3\pi\mu_0\eta D}{\text{vol} \cdot B_s} \int_0^z \frac{-1}{\dfrac{dB(z)}{dz}} dz$$

This equation is for any given magnetic flux density function produced by the magnetic needle, $B(z)$. In the case of magnetic needles, the flux density is being produced by the needle. This flux density is likely to be very complicated and dependent on the geometry of the needle. It would be very difficult to calculate the magnetic flux density for a real needle design without numerical simulations; however, by assuming the needle as an ideal magnetic dipole, a closed-form solution can be determined. The general expression for the magnetic field produced by a magnetic dipole is

$$\mathbf{B}(r) = \frac{\mu_0}{4\pi}\left(\frac{3r(\mathbf{m}\cdot\mathbf{r})}{r^5} - \frac{\mathbf{m}}{r^3}\right)$$

(5.15)

In this equation, $\mathbf{m}$ is the magnetic moment of the needle, in A·m², and r is the position vector originating at the needle. Considering only the on-axis z-dimension, the following simplifications can be made:

$$B_z(z) = \frac{\mu_0}{4\pi}\left(\frac{3z(mz)}{z^5} - \frac{m}{z^3}\right) = \frac{\mu_0}{4\pi}\left(\frac{3m}{z^3} - \frac{m}{z^3}\right) = \frac{\mu_0}{4\pi}\left(\frac{2m}{z^3}\right)$$

$$B_z(z) = \frac{\mu_0}{2\pi}\left(\frac{m}{z^3}\right)$$

(5.16)

Now that an expression has been determined for the magnetic flux density function, Equation 5.16 can be differentiated and substituted into Equation 5.14 to solve for the time of flight:

$$\frac{dB_z(z)}{dz} = \frac{-3m\mu_0}{2\pi}\left(\frac{1}{z^4}\right)$$

$$T(z) = \frac{3\pi\mu_0\eta D}{\text{vol} \cdot B_s} \int_0^z \frac{-1}{\dfrac{dB_z(z)}{dz}} dz$$

$$= \frac{3\pi\mu_0\eta D}{\text{vol} \cdot B_s} \int_0^z \frac{-1}{\dfrac{-3m\mu_0}{2\pi}\left(\dfrac{1}{z^4}\right)} dz$$

$$= \frac{3\pi\mu_0\eta D}{\text{vol} \cdot B_s} \frac{2\pi}{3m\mu_0} \int_0^z z^4 dz$$

$$= \frac{2\pi^2\eta D}{\text{vol} \cdot B_s \cdot m}\left[\frac{1}{5}z^5\right]_0^z$$

$$T(z) = \frac{2\pi^2}{5} \frac{\eta D}{\text{vol} \cdot B_s \cdot m} z^5 \tag{5.17}$$

As a reminder, this equation for time of flight is based on the following assumptions:

- The soft-magnetic particles are fully saturated to a value of B_s, in T.
- The particle is directly on-axis with the needle, so that it only experiences forces in the z-dimension.
- The particles are modeled as magnetic dipoles, magnetized to saturation by the needle.
- The needle is also modeled as a magnetic dipole, producing a known magnetic flux.

Equation 5.17 provides some insights into which variables affect the time of flight. The time of flight increases with viscosity, particle diameter, and the distance from the needle. The time of flight decreases with a larger magnetic volume, higher magnetic saturation of the particle, and a larger magnetic moment produced by the needle. These trends would be expected.

A final simplification can be made by noting that the magnetic volume of the particle can be related to its diameter for a spherical particle:

$$\text{vol} = \psi\frac{4\pi R^3}{3} = \psi\frac{4\pi\left(\dfrac{D}{2}\right)^3}{3} = \psi\frac{4\pi D^3}{24}$$

$$\text{vol} = \psi\frac{\pi D^3}{6} \tag{5.18}$$

In this equation, ψ is the volume ratio of the particle that is magnetic, between 0 and 1. In the case of superparamagnetic nanoparticles, the particles often consist of magnetic materials embedded in a polymer of nonmagnetic material. Since only the magnetic material will experience the magnetic force, the volume ratio, ψ, is important to consider.

By substituting Equation 5.18 into Equation 5.17, the final simplified result for time of flight is found as follows:

$$T(z) = \frac{2\pi^2}{5} \frac{\eta D}{\text{vol} \cdot B_s \cdot m} \cdot z^5 = \frac{2\pi^2}{5} \frac{\eta D}{\psi \frac{\pi D^3}{6} \cdot B_s \cdot m} \cdot z^5$$

$$T(z) = \frac{12\pi}{5} \frac{\eta}{\psi D^2 B_s \cdot m} \cdot z^5 \qquad (5.19)$$

Additionally, Equation 5.19 can be inverted to find the distance of capture for a given duration:

$$z(t) = \left(\frac{5}{12\pi} \cdot \frac{\psi D^2 B_s \cdot m}{\eta} t \right)^{\frac{1}{5}} \qquad (5.20)$$

Equation 5.19 provides additional insight over Equation 5.17. This equation shows that increasing the diameter of the magnetic particles will reduce the time of flight with a second-order relationship. This is because increasing the diameter only increases the drag force in a first-order fashion, but it increases the magnetic volume, and thereby force, in a third-order fashion.

With these new equations, some rough calculations can be made on an actual example for estimation. For simplicity, a cylindrical magnetic needle tip with 1-mm height and 1-mm diameter will be considered. Further, the magnetic flux density of the needle will be assumed to be 1 T, which is a reasonable assumption of the remanent flux density for an Nd–Fe–B magnet. The magnetic moment of this needle is calculated as follows:

$$m = M \cdot \text{vol} = \frac{B}{\mu_0} \cdot \pi R^2 h$$

$$= \frac{1.0}{4\pi \times 10^{-7}} \pi (0.0005)^2 (0.001)$$

$$= 6.25 \times 10^{-4} \text{ A} \cdot \text{m}^2$$

The viscosity of the fluid will be assumed to be 0.001 Pa·s, which is the value for water. The superparamagnetic nanoparticles are assumed to be

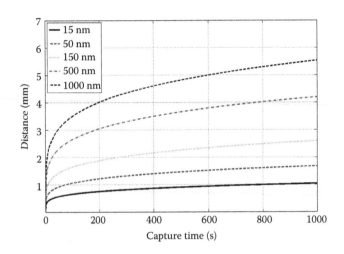

FIGURE 5.17
Plot of distance versus capture time for magnetic nanoparticles of varying diameters.

approximately 50% magnetic by volume and have a saturation flux density of about 0.5 T. Typical diameters for magnetic nanoparticles might range from 15 to 1000 nm. In Figure 5.17, time of flight is plotted against distance from the needle for particles with five different diameters.

It should be noted that these plots assume that the magnetic needle is an ideal magnetic dipole, which is not true. The distance in this plot is actually the distance to the center of the needle tip, which is half a millimeter away from the top edge of the cylinder.

The plots show that increasing the size of the magnetic nanoparticles can have a significant impact on the capture distance for a given time. In 10 minutes, particles with a 15-nm diameter can only be captured from about 1 mm away from the center of the needle. In the same 10-minute timeframe, a 1-μm particle would be captured from 5 mm from the center of the needle.

5.6 Conclusions

Magnetic needles could have multiple usages across various fields. The utility of these needles makes them very attractive. However, understanding the engineering and design considerations behind magnetic needles can greatly improve their efficacy.

Choosing the correct type of needle, material, and geometry can have great impacts on both the practicality and manufacturability of the needle. The advantages and disadvantages of each type of needle have been discussed to provide the reader with a basis for selecting a permanent magnet, electromagnet, or soft magnet-based needle, depending on the restrictions linked to the application. Simulations of each needle type were included for visualization of the magnetic field and the magnetic field gradient distributions. Additionally, the geometry and material considerations have been examined. A brief overview of the manufacturing techniques available for creating small-size magnets was also discussed.

Finally, some applications of magnetic needles were considered, including applying magnetic forces on and capturing magnetic nanoparticles. These applications were accompanied by calculations and theoretical formulas to provide a mathematical basis for understanding how the different variables influence the motion of the magnetic particles and to improve the particle collection with the magnetic needle for a given application.

References

1. O. Cugat, J. Delamare and G. Reyne, Magnetic micro-actuators and systems (MAGMAS), *IEEE Transactions on Magnetics,* vol. 39, no. 5, pp. 3607–3612, Nov. 2003.
2. B. D. Matthews, D. A. LaVan, D. R. Overby, J. Karavitis and D. E. Ingber, Electromagnetic needles with submicron pole tip radii for nanomanipulation, *Applied Physics Letters,* vol. 85, no. 14, pp. 2968–2970, Oct. 2004.
3. D. P. Arnold and N. Wang, Permanent magnets for MEMS, *Journal of Microelectromechanical Systems,* vol. 18, no. 6, pp. 1255–1266, Dec. 2009.
4. B. A. Peterson, F. Herrault, O. D. Oniku, Z. A. Kaufman, D. P. Arnold and M. G. Allen, Assessment of laser-induced damage in laser-micromachined rare-earth, *IEEE Transactions on Magnetics,* vol. 48, no. 11, pp. 3606–3609, Nov. 2012.
5. N. M. Dempsey, Hard magnetic materials for MEMS applications, In *Nanoscale Magnetic Materials and Applications,* New York, NY: Springer, 2009, pp. 661–684.
6. N. Wang, B. J. Bowers and D. P. Arnold, Wax-bonded NdFeB micromagnets for microelectromechanical systems applications, *Journal of Applied Physics,* vol. 103, no. 7, pp. 07E109-1–07E109-3, Apr. 2008.
7. G. T. A. Kovacs, *Micromachined Transducers Sourcebook,* New York, NY: McGraw-Hill, 1998.
8. A. Walther, C. Marcoux, B. Desloges, R. Grechishkin, D. Givord and N. M. Dempsey, Micro-patterning of NdFeB and SmCo magnet films for integration into micro-electro-mechanical-systems, *Journal of Magnetism and Magnetic Materials,* vol. 321, no. 6, pp. 590–594, Mar. 2009.
9. T. H. Boyer, The force on a magnetic dipole, *American Journal of Physics,* vol. 56, no. 8, pp. 688–692, Aug. 1988.

10. L. Vaidman, Torque and force on a magnetic dipole, *American Journal of Physics,* vol. 58, no. 10, pp. 978–983, 1990.

11. A. Garraud, B. Kozissnik, C. Velez, E. G. Yarmola, L. Maldonado-Camrago, C. Rinaldi, K. D. Allen, J. Dobson and D. P. Arnold, Collection of magnetic particles from synovial fluid using Nd-Fe-B micromagnets, In *Symposium of Design, Test, Integration and Packaging of MEMS/MOEMS (DTIP)*, Cannes, France, 2014.

12. A. Garraud, C. Velez, Y. Shah, N. Garraud, B. Kozissnik, E. G. Yarmola, K. D. Allen, J. Dobson and D. P. Arnold, Investigation of the capture of magnetic particles from high-viscosity fluids using permanent magnets, *IEEE Transactions on Biomedical Engineering*, vol. 63, no. 2, pp. 372–378, 2016.

6

Magnetic Delivery of Cell-Based Therapies

Boris Polyak and Richard Sensenig

CONTENTS

6.1 Introduction 123
6.2 Classes and Types of Magnetically Responsive Materials 124
6.3 Principles of Magnetic Cell Delivery and Types of Magnet Systems 127
6.4 Requirements for MNPs and "Therapeutic" Cells 131
6.5 Case Studies.. 135
    6.5.1 Cell Loading 136
    6.5.2 Magnet System 136
6.6 Concluding Remarks.. 139
Acknowledgments.. 140
References.. 141

6.1 Introduction

Cell-based therapy is one of the promising and rapidly growing fields of translational medicine. It stands at the interface of a variety of dynamically developing disciplines, including biomaterials, transplantation, tissue engineering, drug delivery, and stem cell biology. Cell-based therapeutics hold great promise for treating both genetic and acquired diseases. Initially, cell-based therapeutics have been used for blood transfusions and bone marrow transplantations.[1,2] However, recent advances in cell and molecular biology have expanded the potential applications of this approach. Cells are currently used as substitutes for diseased or damaged cells and tissues (cell replacement therapy[3]), components in the reconstruction of regenerated tissues (tissue engineering[4]), and drug delivery vehicles.[5] The use of dendritic cells as cell vaccines is also an emerging immunotherapeutic modality. Dendritic cells can be equipped with antigens and thus can act as cell-based vaccines to prevent tumor relapse.[6,7]

Because cells are self-contained and self-regulated biological entities, they offer several advantages as therapeutic systems. First, cells promise a great deal in the restoration of lost, damaged, or degenerated function, which can be applicable for treatment of cardiovascular diseases, neurological diseases (including Parkinson's, Alzheimer's, and spinal cord injury repair), bone defects, and diabetes. Second, cells can provide paracrine (secretory) effects and regulate the functions of surrounding cells or tissues.[8-11] These effects could be based on cells' nature to secret certain inherent factors, or "therapeutic" cells could be genetically reprogrammed to express and secret factors inducing a regeneration pathway of interest.[12-14] For example, endothelial cells produce a variety of substances in a finely tuned equilibrium, which is crucial for regulation of vascular tone and structure for prevention of cardiovascular diseases such as atherosclerosis.[15] Third, the use of autologous cells eliminates the risk of immunogenicity and does not require implementation of a strategy for engineering of immune acceptance. Fourth, cells themselves can serve as drug delivery vehicles, for example, erythrocyte-based drug delivery.[16,17] These cells can be easily processed and could accommodate both traditional and biologic drugs. Erythrocytes have been evaluated as drug carriers, as they are safe and efficient for treating several diseases. Finally, cells may be minimally manipulated or expanded in cell culture (with or without modification) to obtain a sufficient "cell dose" to provide a therapeutic effect.

To realize the great potential of cell-based therapy, some formidable tasks still have to be addressed. These include achieving immune tolerance (if nonautologous cells used), developing biomimetic materials for tissue engineering, controlling stem cell differentiation, and developing the means of efficient localization of cell-based therapy. One way to specifically localize or target cell-based therapy to a diseased tissue is through generation of magnetic fields and magnetic forces. In this approach, the therapeutic cells have to be loaded with magnetic particles to become responsive to a magnetic force generated by an applied magnetic field gradient. Magnetic particles develop magnetic polarization and magnetophoretic mobility when an external magnetic field and field gradient are applied.[18] Therefore, using the selective application of a magnetic field gradient to the desired area, cells bearing such particles can be successfully carried to the desired site of action with a relatively high accuracy, minimum surgical intervention, and maximized dose. In this way, regional cell therapy efficacy may be improved by increasing local therapeutic cell levels, while systemic "therapy" distribution and undesired side effects may be decreased or eliminated.[19]

This chapter discusses classes and types of magnetically responsive materials, conditions of cell labeling with magnetic particles, fundamental principles of magnetic cell delivery, as well as examines specific cases of the magnetic cell-based therapy approach.

6.2 Classes and Types of Magnetically Responsive Materials

To better understand the fundamental principles of magnetic cell targeting, we first introduce here some basic physical properties of magnetic materials, discuss their behavior in a magnetic field, and identify properties of materials appropriate for targeted cell delivery.

The magnetic properties of a solid originate primarily from electron spin and its associated magnetic moment with a small contribution from the orbital motion.[20] The electrons also determine the strength of the interaction between atoms in a solid, forming the basis for the different macroscopic behavior of materials. At macroscopic levels, the magnetic interactions between the electrons of neighboring atoms, together with the crystalline structure of a solid, determine the magnetic response of materials. In simplified terms, for most materials, the exclusion principle forces adjacent electrons to have opposite spin and therefore opposite magnetic moments which cancel on average. Materials for which every electron with spin +1/2 has an associated spin −1/2 to cancel its magnetic spin moment are called diamagnets and are broadly considered as nonmagnetic, and display a negative, weak linear response to the applied field. Paramagnetic materials are materials that contain unpaired spins and magnetize very weakly along the direction of an external magnetic field. Diamagnetic materials magnetize in the direction opposing the external magnetic field (Figure 6.1a). In the situation of stronger magnetic interactions, the atoms within the solid can align the atomic magnetic moments predominantly parallel (ferromagnet or ferrimagnet) or antiparallel (antiferromagnet) configurations. Ferromagnetic (metals) and ferrimagnetic (metal oxides) materials magnetize strongly along the direction of an external field and tend to retain their magnetization (Figure 6.1b) when the external field is removed,[21] whereas in the antiferromagnets the antiparallel alignment can reduce the total moment to zero, yielding a behavior similar to a paramagnet.

If a macroscopic ferromagnetic material had all its magnetic moments aligned in parallel, a large external magnetic field would be created, which would contain a huge amount of magnetostatic energy. The way in which a solid can reduce this huge magnetostatic energy is to break itself up into regions called magnetic domains. Within a single domain, all magnetic moments remain parallel, but each domain is randomly oriented so that the net magnetic moment of the material is reduced (Figure 6.1c). This situation generates interfaces between domains called domain walls (DWs), where adjacent magnetic moments are in a nonfavorable configuration so that these DWs are highly energetic areas. Even though some energy is stored inside DWs, the overall decrease in the total magnetic energy favors the material adopting a multidomain configuration. The bulk material is split into more domains until the energy associated with one more wall is as large as the energy decrease in the magnetic field outside the sample.[22]

Upon magnetization, ferro- and ferrimagnetic materials have remnant magnetization (remanence in magnetic literature, M_R) even when the field is removed (Figure 6.1b), and hence they tend to interact magnetically and stick to each other, forming aggregates. These aggregates can cause accidental embolisms if they enter the bloodstream or other permanent magnetic-related

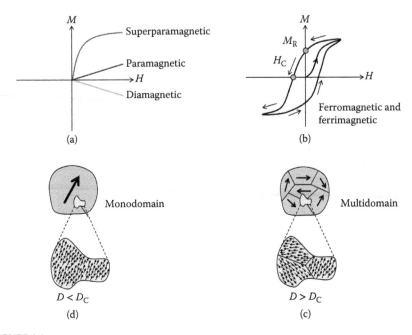

FIGURE 6.1
Materials show a wide range of magnetic behavior. Magnetic moment, M as function of a magnetic field, H of: (a) paramagnets, diamagnets, and superparamagnets and (b) ferro- or ferrimagnets. M_R is remnant (residual) magnetization preserved in ferro- and ferrimagnets when an external magnetic field is removed. H_C is the coercive force measured by intensity of the applied magnetic field required to reduce the magnetization of the material to zero after the magnetization of the sample has been driven to saturation. (c) Schematic view of magnetic domains in a multidomain ferromagnetic particle having size larger than the critical diameter $D > D_C$. For this particle, the whole material breaks down into randomly oriented magnetic regions (domains) reducing the total magnetostatic energy. (d) For $D < D_C$, the material becomes a single-domain particle behaving as superparamagnetic material.

interferences, and therefore, ferro- and ferrimagnetic particles are generally avoided in biomedical applications—at least in sizes where they are able to preserve a stable magnetization in the absence of a magnetic field.

The field energy associated with a ferromagnet or ferromagnetic domain is related to the volume of the domain, whereas the DW energy is related to the surface area of the wall. As the volume of a bulk ferro- or ferrimagnetic material is reduced, and the domains reduce in size, the volume decreases more rapidly than the surface area until the volume reaches a certain value, called critical domain size, D_C, where maintaining a DW is no longer an energetically favorable state. Consequently, the material adopts a single-domain configuration (Figure 6.1d). Within this single magnetic domain, all the atomic magnetic moments aligned along the same direction. A single domain particle can remain a ferromagnet. However, at such small sizes, in the absence of a magnetic field, thermal energy can force the magnetic moments to flip rapidly. This "flipping" is very rapid (on the order of 10^{-9}/sec), resulting in a total zero average magnetic moment at room temperature during the time of the measurement (Figure 6.1a) forming a superparamagnet.[23] The value of critical size D_C below which a particle of a given material becomes single domain is determined by intrinsic properties of each material (e.g., magnetic anisotropy, magnetic moment, and exchange anisotropy) and also dependent on the particle shape. Single-domain critical sizes D_C for spherical metallic iron particles range about 7–11 nm and for the iron oxide (magnetite, Fe_3O_4) are about 20–30 nm.

In biomedical applications, superparamagnetic particles are typically chosen as a compromise between the desire to achieve strong magnetization and the need to avoid permanently magnetized objects. Magnetite (Fe_3O_4) and its oxidation product maghemite (γ-Fe_2O_3) are the two most widely used magnetic materials in the biomedical field. These materials fulfill the requirements of high saturation magnetization at room temperature (M_S ~ 92 emu/g for bulk magnetite and 50–65 emu/g for various types of magnetite nanoparticles[24] and 73–74 emu/g for bulk maghemite[25] and 7–65 emu/g for various types of maghemite nanoparticles[26]) and have the lowest toxicity levels yet known in preclinical tests.[27,28]

6.3 Principles of Magnetic Cell Delivery and Types of Magnet Systems

Magnetic cell delivery comprises three basic components: (1) "therapeutic" cells capable of regenerating damaged cells or tissue through their intrinsic functional properties or via genetic encoding to produce regulatory biomolecules; (2) superparamagnetic carrier particles, which are usually loaded within cells or attached to a cellular membrane rendering cells magnetically responsive; and (3) a magnet system providing both a source of magnetic field to transiently magnetize particles associated with cells and a source of field gradient generating pulling force to draw magnetically responsive cells toward target tissues.

In magnetic cell delivery, the superparamagnetic particles are carriers of cells, and therefore we will consider cells associated with such particles as magnetizable and magnetically responsive objects. The force acting on such magnetically responsive cells with magnetic moment $\vec{M}$ is governed by the equation $\vec{F} = (\vec{M} \cdot \nabla \vec{B}) \vec{B}$. It is clear that in order to maximize the force acting on a magnetically responsive cell, the magnet system has to generate sufficiently strong field $\vec{B}$ to maximize the cell's induced magnetization $\vec{M}$ as well as create strong field gradients at the cell's location to pull cells toward the target. For physical reasons, no single source of a magnetic field can fulfill both conditions.[19] Depending on the targeting application, additional physical forces opposing cell targeting should be considered in the design of the magnet system. For example, in the vascular cell targeting application, the main force that needs to be overcome for the targeting of suspended cells is the drag force (i.e., Stokes drag force) of the fluid defined by the equation $F_d = -6\pi\eta R_c(v_c - v_f)$, where R_c is the cell radius, η is the dynamic viscosity, and is the $v_c - v_f$ velocity difference between cell and fluid. The Stokes drag force is only accurate for low Reynolds numbers as commonly found for magnetic cell targeting in vascular applications.

The magnet system used to attract and localize magnetic nanoparticle (MNP)-loaded cells to the desired target is an important, but often overlooked, component of the overall cell targeting system. In most systems, the field generated by the magnet component must perform the functions of both magnetizing the MNP-loaded cells and generating an attractive force to the target area. In general, it is difficult to optimize both these functions with a single design because the magnetizing function depends on field strength whereas the force is a function of the field gradient.

Many studies have used strong external magnets placed close to the target area. This approach often yields good results in small animals but cannot be scaled up to large animals or human dimensions.[19] Targeting systems for practical applications must be designed in such a way that the required field and field gradients can be induced at the targeting site in a clinical setting. Several approaches have been used for implementing the targeting magnet system in animal models: placement of permanent magnets exterior to the targeting site,[29–35] implantation of permanent magnets,[36–39] external electromagnets,[40–43] external magnet arrays,[44] and magnetizable implants.[45,46] In addition to these studies, a substantial body of literature has been published on magnet systems for targeted drug delivery, which has direct applicability to cell delivery.[47]

The most common and simplest magnet system is the use of external permanent or electromagnets. External permanent magnets can be a good choice for initial studies in which proof of concept is the endpoint. Rare-earth magnets of great variety in shape and magnetic characteristics are easily available commercially. Electromagnets can generate strong fields and can provide the capability of modifying the field strength or even be switched off throughout an experiment if desired. Electromagnets are subjected to heating losses, and care must be taken to keep the surface temperature below approximately 40°C near the test subject.[47] The use of external magnets, however, imposes serious limitations in targeting deep tissues as their field strength and field gradient decrease exponentially with distance from their surface as further discussed in Section 6.5.2. The implanted permanent magnets have also drawbacks as they require invasive surgical procedures for placement and extraction, pose the risk of a permanent magnetic field within the body, and they can move or interfere with a magnetic resonance imaging (MRI) procedure.

In the following example, we compare magnetic properties of two cylindrical NdFeB permanent magnets with the same thickness of 19.0 mm but with different pole diameters of 1.59 mm, that is, "narrow" magnet and 12.7 mm, that is, "wide" magnet. Figure 6.2 shows the magnetic flux density and field gradients of both magnets as a function of distance from the magnets' pole surface.

At the distance of the magnet's half diameter, both magnets generate comparable magnetic flux density (~0.2 T) sufficient to magnetize magnetically labeled cells reaching their saturation magnetization limit. However, at this distance, the field gradient of the "narrow" magnet is an order of magnitude higher (~400 T/m) compared to the "wide" magnet (~40 T/m), Figure 6.2a. The magnitude of such high field gradient of the "narrow" magnet would be sufficient to generate strong pulling forces to target cells, but the effective working distance of such magnet is limited to less than 1 mm (Figure 6.2a).

At distances larger than a diameter of the "narrow" magnet, its magnetic flux density and field gradient drop exponentially to a noneffective range, which altogether makes such a magnet impractical for targeting applications. In contrast, the magnetic flux density of the "wide" magnet drops with distance moderately, resulting in lower field gradients and hence significantly weaker magnetic forces applied to the cells. At the distance of its diameter (~13 mm), this magnet would generate a considerably low gradient (~10 T/m) and low magnetic flux density (~0.06 T), Figure 6.2b. Both values are too weak to enable sufficient magnetization and generation of pulling forces acting on the therapeutic cells in the bloodstream. This example demonstrates that systems based on permanent magnets are unable to provide

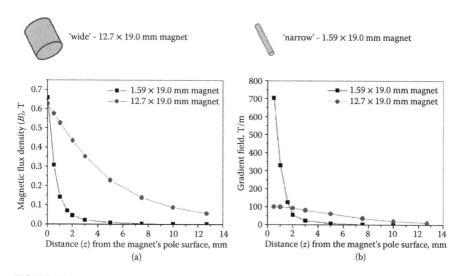

FIGURE 6.2
Magnetic properties of two cylindrical NdFeB permanent magnets with the same thickness of 19.0 mm but with different pole diameters of 1.59 and 12.7 mm. (a) magnetic flux density and (b) field gradients of both magnets as function of distance from the magnets' pole surface.

both sufficiently strong magnetizing field and strong field gradient at distances relevant to human scale.

Magnet arrays provide the potential for increasing the field in the target area while reducing it elsewhere. The most common magnet array is the Halbach array,[48] which augments the magnetic field on one side of the array while canceling the field to near zero on the other side. Considerable work has been done with these arrays in applications ranging from particle accelerators, where they were first used, to magnetic levitation trains. However, their application to cell targeting has been utilized only to a minor extent, and the potential of this approach remains mostly unrealized. An application using the approach discussed in Section 6.5.

Magnetizable implants, such as a stent, can be used to partially overcome the limitation of optimizing both magnetizing the MNP-loaded cells and generating an attractive force to the target area discussed above. In the concept developed by Polyak and coworkers[46] as shown in Figure 6.3, a steel stent wire stent implanted in a blood vessel.

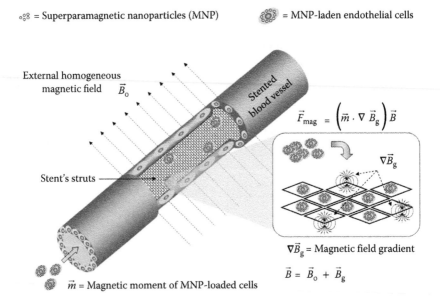

FIGURE 6.3

A schematic representation of a stented blood vessel showing that magnetically responsive cells are attracted to steel stent struts in a uniform, magnetic field because of a generated magnetic force ($\vec{F}_{mag}$) that directly depends on the strength of the total magnetic field ($\vec{B}$), high field magnetic gradients ($\nabla \vec{B}_g$) induced on the stent struts, and the magnetic moments ($\vec{m}$) induced on MNP-loaded cells by the uniform magnetic field $\vec{B}_o$. The total magnetic field ($\vec{B}$) is a sum of a gradient field ($\vec{B}_g$) of the stent and a uniform field ($\vec{B}_o$). The steel mesh structure of the stent is shown in both the open stented blood vessel fragment and as individual strut areas in inset. (Reprinted from Polyak, B. et al., *Proc Natl Acad Sci U S A*, 105, 698–703, 2008. With permission.)

A modest external uniform magnetic field (~0.2 T) both magnetizes the MNP-loaded cells and produces large local magnetic field gradients within the steel stent wire network. The generation of such modest uniform magnetic field does not require sophisticated magnetic setups, which can allow easy transition of the magnetic targeting technique to larger animals and humans. The magnetizable stent used in this approach was made of 304-grade stainless-steel and exhibited near superparamagnetic behavior showing slight hysteresis and a remnant magnetization on the order of 7% of its saturation magnetization value (~9.8 emu/g).[46] The application of this technique is discussed in Section 6.5.

6.4 Requirements for MNPs and "Therapeutic" Cells

MNPs and "therapeutic" cells need to possess certain properties to enable successful cell targeting. The MNPs have to be nontoxic, biocompatible, and inert regarding influencing cell fate (e.g., unable to affect the "stemness" of undifferentiated cells or incapable of changing functional properties or dedifferentiating specialized cells), able to strongly magnetize, and enable safe and easy protocols to achieve efficient cell loadings. The "therapeutic" cells have to be able to efficiently internalize MNPs while preserving their "stemness" or other functional properties and showing no or minimal toxicity effects.

As was discussed earlier, magnetite (Fe_3O_4) or maghemite (γ-Fe_2O_3) nanoparticles offer sufficiently good magnetic properties balanced by low toxicity. To enhance the cumulative magnetic responsiveness of magnetite nanoparticles per volume unit, they could be entrapped within a hydrophobic polymer matrix using biocompatible and biodegradable polymers, that is, poly(lactic acid), PLA or poly(lactic-co-glycolic acid), PLGA[46,49] or stabilized with a hydrophilic polymer, for example dextran.[50] Such polymer-based particles enable incorporation of up to 50% (w/w) magnetite within the polymer matrix and have a saturation magnetization of ~25 emu/g MNP composite (Figure 6.4).

An additional advantage of such polymer-based MNPs is their submicron size with a mean diameter of 300–400 nm. MNPs of this size range experience rapid (up to 10 minutes) magnetic sedimentation and become available at high concentrations for uptake at the cell membrane surface.

Another interesting observation related to such polymer-based MNPs is their enhanced internalization under the influence of a magnetic force as demonstrated in endothelial cells, relevant to vascular cell targeting applications. It has been shown that the internalization rate of MNPs by endothelial cells is dependent on the magnetite loading within particles, indicating that MNP uptake is a force-dependent process (Figure 6.5a). Also,

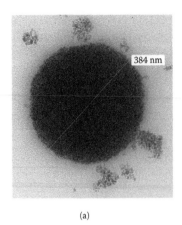
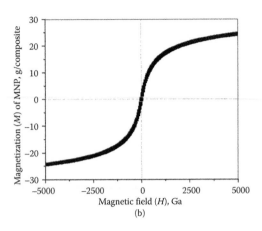

(a)   (b)

FIGURE 6.4
Characterization of magnetic nanoparticles (MNPs). (a) Transmission electron micrograph of 50% (w/w) MNP. (b) Magnetization curve of 50% (w/w) MNP recorded at room temperature by an alternating gradient magnetometer.

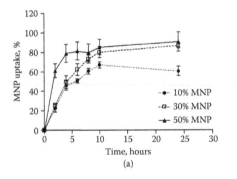
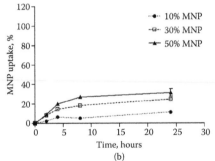

(a)   (b)

FIGURE 6.5
The kinetics of 10%, 30%, and 50% (w/w) MNP uptake by bovine aortic endothelial cells (BAECs) in culture as a function of incubation time and various magnetic field conditions. (a) A magnetic field of 500 G "On" by using a fixed magnet applied directly to the underside of the cell culture plates. (b) Magnetic field "Off." (Adapted with permission from **Springer Science+Business Media**: *Pharm Res*, 29, 2012, 1270–81, MacDonald, C. et al., copyright © 2012, Springer Science+Business Media, LLC.)

the degree of MNP internalization in the absence of a magnetic field is dramatically reduced (Figure 6.5b).[51] This study demonstrates that endothelial cells could be quickly and efficiently loaded with 50% (w/w) MNPs within 4 hours enabling a high magnetite content of ~20–25 pg/cell. This MNP loading results in a saturation magnetization of 400 nemu/cell. It is of note that such high MNP-loading of endothelial cells did not markedly affect cellular metabolic activity, ability to proliferate, cell functional competence, gene expression, intracellular organelles, or cell respiration as compared to control nonloaded cells.[51-53]

Endothelial cells are known as highly phagocytic cells[54] and therefore achieving high MNP contents within these cells is a relatively simple task. In contrast, labeling other nonphagocytotic cell types relevant to the therapeutic application (e.g., mesenchymal stem cells, MSCs; neural stem cells, NSCs; hematopoietic stem cells, etc.) is a much more challenging objective as these cells usually lack particular endocytosis mechanisms. Various stem cell types have been shown to internalize MNPs at the level of 2–10 pg iron/cell[55-57] utilizing different strategies. These include the use of extended particle incubations with cells (up to 48 hours),[58] utilization of chemical transfection agents (such as chitosan or polylysine),[59] or use of cell uptake enhancing molecules (such as the RGD and TAT peptides).[60] However, these strategies are usually time-consuming and can be associated with cellular toxicity and changes in cellular physiology.[60,61] Therefore, the need still exists for the development of safe and effective nonchemical means of enabling the loading of stem cells with MNPs to sufficient levels for cell targeting applications.

Another important aspect in the preparation of "therapeutic" cells for magnetic cell delivery is the preservation of their functionality or "stemness." For example, human mesenchymal stem cells (hMSCs) are good candidates for cellular therapies because their paracrine function can modulate in response to the microenvironment in a therapeutic manner and because they can differentiate into various cell lineages, which make them suitable for cell replacement therapy.[62,63] The preservation of "stemness" has been demonstrated in a recent study in which hMSCs were labeled with superparamagnetic iron oxide nanoparticles (SPIOs).[57] In this study, the cell viability was determined immediately after labeling or 7 days later. In neither case did the authors observe a toxic effect in the studied nanoparticle concentration range. To determine the extent to which SPIO labeling altered stem cell characteristics in hMSCs, the expression of key surface markers in unlabeled and labeled cells was evaluated. The authors found that SPIO labeling did not alter the immunophenotype of the cells. Furthermore, labeled and unlabeled cells were cultured at conditions that induce differentiation of MSCs into adipogenic, osteogenic, and chondrogenic lineages. Cells differentiated into these lineages in a similar manner regardless of the presence of SPIOs. Interestingly, SPIOs remained present in labeled cells after differentiation.[57]

In another study, Chari and coworkers have demonstrated that increased nanoparticle magnetic content in conjunction with applied magnetic fields (static and oscillating) safely improves labeling of primary NSCs—"hard-to-label" cell transplant population without compromising their "stemness" and differentiation potential (Figure 6.6).[64]

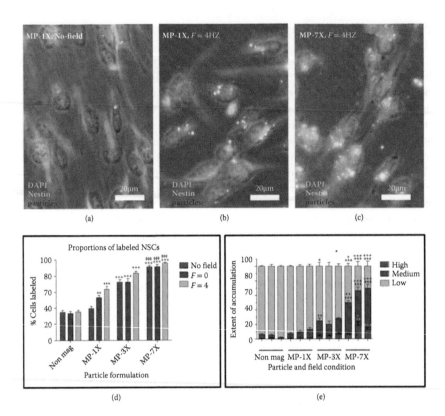

FIGURE 6.6
Assessment of cellular uptake in NSCs when using the test particle formulations under different magnetic fields. Representative triple merged images of NSCs labeled with MP-1X under (a) no field and (b) $F = 4$ Hz magnetic fields and (c) MP-5X under $F = 4$ Hz magnetic field. Note that greater proportions of cells appear to be labeled under application of a 4 Hz magnetic field when using the MP-1X formulation. (d) Bar chart displaying quantification of percentage of NSCs labeled with the various particle formulations under different magnetic fields. Statistical differences are *$p < .05$ and ***$p < .001$ versus no field condition labeling at the same particle iron concentration; +++$p < .001$ versus MP-1X labeling under the same field condition; ⊖⊖$p < .01$ and ⊖⊖⊖$p < .001$ versus MP-3X labeling under the same field condition. (e) Bar chart displaying semiquantitative analysis of extent of particle uptake using the various particle formulations under the different magnetic field conditions. Comparable low, medium, and high groups were analyzed for statistical differences and are *$p < .05$, **$p < .01$, and ***$p < .001$ versus MP-1X labeling under the same field condition; +$p < .05$, ++$p < .01$, and +++p .001 versus MP-3X labeling under the same field condition (One way ANOVA with Bonferroni's MCT, $n = 5$). (Adapted from Elsevier, Vol. 11, Adams, C. F. et al. Increasing magnetite contents of polymeric magnetic particles dramatically improves labeling of neural stem cell transplant populations, pp. 19–29, 2015, Copyright [2015], with permission from **Elsevier**.)

6.5 Case Studies

The magnetic delivery of cell-based therapies has been applied to a variety of potential clinical applications. At this time, however, none of these efforts have advanced past the animal model experimental stage. Applications include retinal degeneration,[35] spinal cord injury,[38–40,65] orthopedics,[37,41,42] vascular,[32,44–46,66–69], and cell delivery to and accumulation in major organs.[30,33,36,70] There is no doubt that other applications will be investigated and developed as the technology matures.

In this section, we discuss a magnetic cell delivery application in the context of treatment of coronary and peripheral artery disease. Angioplasty has become the most common revascularization procedure for coronary and peripheral artery disease. However, recurrent arterial narrowing (restenosis) is the major complication limiting the success of this revascularization procedure. A combination of processes causes restenosis. In the short term, it may be caused by elastic recoil of the vessel wall, thrombus formation at the site of injury, and variations in operative technique that lead to a smaller anastomosis or kinking of the vessel. Longer term patency over the preceding months to years may be limited by intimal hyperplasia, involving the proliferation and migration of intimal smooth muscle cells (SMCs).[71,72] The overall incidence or restenosis is approximately 30% a year after coronary angioplasty and bare metal stenting,[73] and there is a similar incidence following angioplasty for peripheral arterial disease.[74] Drug-eluting stents have been developed as a means of preventing intimal hyperplasia and appear to have reduced the early risk of coronary restenosis, although this still occurs in over 10% of stented vessels at 12 months.[73] There are also concerns of an increased incidence of stent thrombosis and myocardial infarction in patients who have had a drug-eluting stent inserted.[73]

Preventing intimal hyperplasia is an important therapeutic target and strategies include not only continued development of stent design and coating materials,[75,76] but also manipulation of the cellular response to vascular injury. In addition to regulating vascular tone and permeability, the endothelium is emerging as a key modulator of cellular response to vessel wall injury.[77,78] Accelerated reendothelialization following arterial injury inhibits SMC proliferation[79] and therefore has the potential to prevent or reduce intimal hyperplasia.[80] Early restoration of endothelial integrity has been shown to modify thrombogenic and proliferative vascular wall properties in several animal models.[81–85] Clinical observations and animal studies have led investigators to test the hypothesis that enhancing reendothelialization might inhibit neointimal formation following vascular injury. Indeed, early restoration of endothelial integrity through local vascular endothelial growth factor administration was associated with reduced proliferation of intimal vascular SMCs, providing indirect evidence to support a role for endothelial integrity in suppressing intimal hyperplasia.[86–88]

Several groups have applied magnetic delivery of endothelial cells to stented[45,46,66,68] or nonstented[44,69] blood vessels, synthetic vascular magnetic grafts[89] or stent-grafts[67] to assess the feasibility and efficacy of this approach. In all these studies, the endothelial cells are loaded with MNPs and administered in the vicinity of the stent, vascular graft, or targeted vessel region. All the groups demonstrated a level of success in the localization of endothelial cells to stents and targeted vessel regions in animal models. Nevertheless, these groups differ in their approaches, particularly in the design of the targeting magnet system. These approaches serve as good examples of the types of trade-offs that must be made in implementing an *in vivo* targeting system.

6.5.1 Cell Loading

The above groups took different approaches in loading the cells with MNPs, both of which were effective in achieving adequate loading. In several studies, the desired cell loading levels were achieved by incubating the endothelial cells with MNPs overnight without employing any active mechanisms.[45,69] One study reported that pretreatment of endothelial cells for 2 hours with serum-free medium before nanoparticle addition could lead to increased particle internalization.[44] Alternatively, a magnetic force-mediated internalization achieved by exposing the cells incubated with MNPs to a magnetic field of 500 G and surface force density of 66 T^2/m (LifeSep 96F, Dexter Magnetic Technologies, USA).[46,66,68] In this case, the high levels of MNP loading (~25 pg magnetite/cell) can be reached as soon as 4 hours without compromising endothelial cell well-being and functionality.[51–53] Both approaches were effective is likely because endothelial cells readily internalize MNPs owing to their phagocytic nature. Other cell types could require an active approach.

6.5.2 Magnet System

Targeting blood vessels without metallic stents requires the use of external magnets. Vosen and coworkers used an arrangement of twelve bar magnets to generate radially symmetric magnetic field.[69] This strategy resulted in circumferential retention of MNP-loaded eNOS-overexpressing endothelial cells at the wall of mechanically injured vessels *ex vivo* and *in vivo*, yielding an improvement of vascular function in a murine model of vascular injury.[69] Similarly, a cylindrical Halbach array was used to target MSCs in a rabbit model of vascular injury.[44] This study demonstrated a sixfold increase in cell retention leading to a nonsignificant reduction of restenosis 3 weeks after magnetic cell delivery.[44] While those studies demonstrated the potential of magnetic cell delivery for vascular applications using external magnets, scaling up magnet systems based on this design to humans would be a significant challenge due to the physical limitations discussed in Section 6.3.

Magnetizable implants such as stents represent a more realistic approach to magnetic cell delivery because of scalability of magnet systems to human dimensions. Using premagnetized stents as was described by Pislaru and coworkers does not require any magnet systems for the realization of targeting at the time of the cell delivery procedure.[45] In this approach, a stent could be premagnetized before implantation. This design process, however, requires a trade-off in selecting the stent material between the standard structural properties required for the stent and the additional magnetic properties required for attracting the MNP-loaded endothelial cells. Through experiment, it was determined that commercially available stainless steel stents with a 10-μm-thick layer of nickel were the best compromise solution. The nickel-coated stainless steel stents were brought into contact with a 5000 G neodymium-iron-boron magnet for 5 minutes immediately before implantation in a porcine model. No external field was applied to the animal, so the stent itself induced the magnetization of the cells and generated pulling magnetic forces to capture cells on the stent surface. One of the disadvantages of using a permanently magnetized implant as described in this example is potential problems with later MRI studies. However, the authors stated that the magnetized stents in their approach are likely to create localized imaging artifacts if an individual requires an MRI, but are unlikely to result in contraindication of future imaging. Additionally, nickel coating is expected to induce local toxic effects due to oxidation of the metal coating and release of nickel ions (capable of inducing allergic reactions) in the vicinity of the stent over time. The long-term magnetic properties of the magnetized stent were not evaluated but surely degrade over time due to oxidation of the nickel coating and its dissolution. The inability to remagnetize the stent later if redosing is required could be a limitation of this approach. An advantage of this approach is that the problems associated with designing an external magnet configuration whose geometry is compatible with potential patients of varying size while maintaining the proper magnetic field characteristics at the target are removed from the overall design problem.

The magnet system used by the Polyak group is described in Section 6.3 and depicted in Figure 6.3.[46] The stent material used was 304-grade stainless steel chosen for its combination of suitable magnetic properties and its corrosion resistance in aqueous environments. Other alloy steels were evaluated as part of the design process. The main advantage of this approach is that a modest externally applied field can both magnetize the MNP-loaded cells and produce large local magnetic field gradients within the steel stent wire network maximizing the fraction of captured MNP-loaded cells. Since the stent does not retain significant magnetization after the external field is removed, any problems of retaining a permanent magnet within the patient are mitigated. If redosing is required, the stent can be remagnetized through the application of an external field. Using this approach, a recent study demonstrated that magnetic endothelial cell delivery to transiently magnetized stents in a rat carotid artery stenting model enhanced capture, retention, and proliferation of targeted cells at the site of stent implantation.[66] Another recent study showed that magnetically mediated targeting of nonmodified syngeneic endothelial cells prevented the development of in-stent stenosis nearly twofold earlier and with a twofold greater magnitude in treated animals in comparison to untreated controls in a rat carotid artery stent angioplasty model, which was confirmed by both ultrasound (Figure 6.7) and endpoint morphometric analysis (Figure 6.8).[68]

Both of these approaches require an invasive procedure to implant the stent. This is not a problem for the angioplasty application since a stent of some type would be implanted regardless of whether cell therapy was used

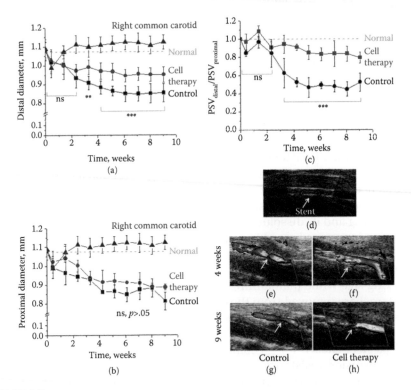

FIGURE 6.7
The protection from in-stent stenosis by the magnetically targeted EC to the stented arteries assessed by the ultrasound. (a and b) Changes in diameter of the stented artery at the distal and proximal end, respectively ($n \geq 11$). (c) Ratios of the peak systolic velocity (PSV) at the distal to proximal ends in studied animal groups ($n \geq 11$). (d) B-mode ultrasound image of the stented left carotid artery. (e through h) Representative color Doppler images of the stented arteries at different time points. Arrows indicate distal end of the stented artery segment. Data represent the means ±SD. Data comparisons were made using one-way ANOVA with Tukey's posthoc test. $^{**}p < .01$ and $^{***}p < .001$ versus untreated (control) arteries; ns, nonsignificant, $p > .05$. Normal are the values measured prior to stent implantation. (Reprinted with permission from [Polyak et al., 2016, 9559–9569]. Copyright [2016] **American Chemical Society**.)

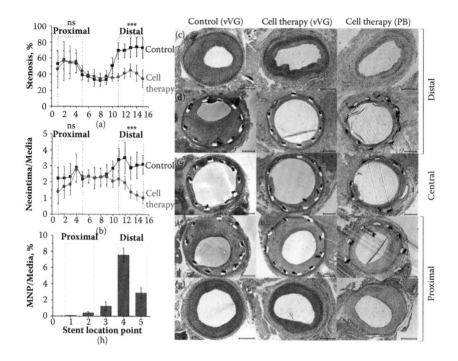

FIGURE 6.8
The protection from in-stent stenosis by the magnetically targeted EC to the stented arteries assessed by morphometric analysis. The results expressed as (a) percent of stent stenosis and (b) neointima/media ratios along the stent ($n \geq 11$). (c through g) Representative Verhoeff-van Gieson (vVG) and Prussian blue (PB) with nuclear fast red as counterstain sections of stented arteries in control and cell therapy group at different stent location points, respectively. Blue staining in the Prussian blue-stained histological slices indicate presence of MNPs in the vessel wall. (h) Distribution of the MNPs along the stented artery segment quantified from the Prussian blue-stained histological slices. The scale is 0.25 mm. Data represent the means ±SD. Data comparisons were made using two-tailed unpaired Student's *t*-test. ***$p < .001$ versus untreated (control) arteries; ns, nonsignificant, $p > .05$. (Reprinted with permission from [Polyak et al., 2016, 9559–9569]. Copyright [2016] **American Chemical Society**.)

or not. However, in other applications, where no invasive procedure is part of standard care, the requirement to install an implant could make other approaches a more favorable option.

6.6 Concluding Remarks

Site-specific delivery of "therapeutic" cells utilizing magnetic targeting strategies has considerable potential for cell and tissue regeneration applications. However, several challenges still have to be addressed for successful translation of the magnetic cell targeting methodology into the clinic.

First, the magnetic particle carriers should be further optimized to possess the highest possible magnetization while enabling fast and safe protocols for efficient cell loading without compromising cellular viability, function, and "stemness" (in the case of undifferentiated stem cells). Further studies elucidating the long-term effects of magnetic carriers on cellular physiology and function are needed to confirm MNP indifference in regard to cellular fate.

Second, there is a need to develop an efficient means to achieve cell loading with MNPs sufficient for targeting and MRI tracking, especially in "hard-to-load" stem cells, the main future therapeutic cell population for tissue regeneration. Such strategies, for example, may include the application of magnetic field gradients in combination with oscillating magnetic fields as has been shown to enhance transfection in "hard-to-transfect" neural cells.[90,91]

Third, more attention is needed for the design of magnet systems enabling generation of sufficiently strong magnetizing fields along with considerably high field gradients at the target tissue location. While many small animal studies have successfully demonstrated the feasibility of magnetic cell targeting using a single source of magnetic field, these systems are not practical for human-scale clinical applications due to fundamental physical reasons. In contrast, use of magnetizable implants (in particular intravascular stents) in conjunction with a modest uniform magnetic field for generation of both high field gradients at the target site and sufficiently strong magnetizing field seem to be feasible and scalable to human dimensions and hence has the greatest potential to be translated into clinical settings. In this regard, the chemical composition of implant material and its geometry may play a key role in trading off the best magnetic properties and biocompatibility for achieving efficient magnetic cell targeting.

Finally, future preclinical studies have to focus on the therapeutic outcomes of the magnetic cell delivery in attenuating target pathological conditions. For example, in vascular applications, the regional delivery of mature endothelial cells or endothelial precursor cells to stented blood vessels is expected to reduce or prevent the formation of neointimal hyperplasia through induced reendothelialization of the stent and the surrounding tissue. It is also clear that successful translation of magnetic cell-based therapy to the clinic will require the joint efforts of researchers from multiple disciplinary backgrounds to overcome the above challenges.

Acknowledgments

This work was supported by the NIH Award R01HL107771, W. W. Smith Charitable Trust Award H1504, Drexel University College of Medicine Clinical & Translational Research Institute (CTRI), and 2016 Commonwealth Universal Research Enhancement (CURE) grant.

References

1. Hodby, K., Pamphilon, D. Concise review: Expanding roles for hematopoietic cellular therapy and the blood transfusion services. *Stem Cells* 2011, 29, 1322–6.
2. Loren, A. W., Porter, D. L. Donor leukocyte infusions for the treatment of relapsed acute leukemia after allogeneic stem cell transplantation. *Bone Marrow Transplant* 2008, 41, 483–93.
3. Mountford, J. C. Human embryonic stem cells: Origins, characteristics and potential for regenerative therapy. *Transfus Med* 2008, 18, 1–12.
4. Cezar, C. A., Mooney, D. J. Biomaterial-based delivery for skeletal muscle repair. *Adv Drug Deliv Rev* 2015, 84, 188–97.
5. Hastings, C. L., Roche, E. T., Ruiz-Hernandez, E., Schenke-Layland, K., Walsh, C. J., Duffy, G. P. Drug and cell delivery for cardiac regeneration. *Adv Drug Deliv Rev* 2015, 84, 85–106.
6. Anguille, S., Smits, E. L., Lion, E., van Tendeloo, V. F., Berneman, Z. N. Clinical use of dendritic cells for cancer therapy. *Lancet Oncol* 2014, 15, e257–67.
7. Gelao, L., Criscitiello, C., Esposito, A., De Laurentiis, M., Fumagalli, L., Locatelli, M. A., Minchella, I., Santangelo, M., De Placido, S., Goldhirsch, A., Curigliano, G. Dendritic cell-based vaccines: Clinical applications in breast cancer. *Immunotherapy* 2014, 6, 349–60.
8. Liu, G. S., Peshavariya, H. M., Higuchi, M., Chan, E. C., Dusting, G. J., Jiang, F. Pharmacological priming of adipose-derived stem cells for paracrine VEGF production with deferoxamine. *J Tissue Eng Regen Med* 2016, 10, E167–76.
9. Lynch, K. M., Ahsan, T. Correlating the effects of bone morphogenic protein to secreted soluble factors from fibroblasts and mesenchymal stem cells in regulating regenerative processes *in vitro. Tissue Eng Part A* 2014, 20, 3122–9.
10. Masuda, H., Kalka, C., Asahara, T. Endothelial progenitor cells for regeneration. *Hum Cell* 2000, 13, 153–60.
11. Matsuda, K., Falkenberg, K. J., Woods, A. A., Choi, Y. S., Morrison, W. A., Dilley, R. J. Adipose-derived stem cells promote angiogenesis and tissue formation for in vivo tissue engineering. *Tissue Eng Part A* 2013, 19, 1327–35.
12. Haastert, K., Mauritz, C., Matthies, C., Grothe, C. Autologous adult human Schwann cells genetically modified to provide alternative cellular transplants in peripheral nerve regeneration. *J Neurosurg* 2006, 104, 778–86.
13. Kumagai, G., Tsoulfas, P., Toh, S., McNiece, I., Bramlett, H. M., Dietrich, W. D. Genetically modified mesenchymal stem cells (MSCs) promote axonal regeneration and prevent hypersensitivity after spinal cord injury. *Exp Neurol* 2013, 248, 369–80.
14. Lin, C. Y., Chang, Y. H., Kao, C. Y., Lu, C. H., Sung, L. Y., Yen, T. C., Lin, K. J., Hu, Y. C. Augmented healing of critical-size calvarial defects by baculovirus-engineered MSCs that persistently express growth factors. *Biomaterials* 2012, 33, 3682–92.
15. Bonetti, P. O., Lerman, L. O., Lerman, A. Endothelial dysfunction: A marker of atherosclerotic risk. *Arterioscler Thromb Vasc Biol* 2003, 23, 168–75.
16. Magnani, M. Erythrocytes as carriers for drugs: The transition from the laboratory to the clinic is approaching. *Expert Opin Biol Ther* 2012, 12, 137–8.

17. Rossi, L., Serafini, S., Pierige, F., Castro, M., Ambrosini, M. I., Knafelz, D., Damonte, G., Annese, V., Latiano, A., Bossa, F., Magnani, M. Erythrocytes as a controlled drug delivery system: Clinical evidences. *J Control Release* 2006, 116, e43–5.

18. Alexiou, C., Arnold, W., Klein, R. J., Parak, F. G., Hulin, P., Bergemann, C., Erhardt, W., Wagenpfeil, S., Lubbe, A. S. Locoregional cancer treatment with magnetic drug targeting. *Cancer Res* 2000, 60, 6641–8.

19. Polyak, B., Friedman, G. Magnetic targeting for site-specific drug delivery: Applications and clinical potential. *Expert Opin Drug Deliv* 2009, 6, 53–70.

20. Martin, D. H. *Magnetism in Solids*, Cambridge, MA: The MIT Press, 1967.

21. Olsvik, O., Popovic, T., Skjerve, E., Cudjoe, K. S., Hornes, E., Ugelstad, J., Uhlen, M. Magnetic separation techniques in diagnostic microbiology. *Clin Microbiol Rev* 1994, 7, 43–54.

22. Feynman, R. P., Leighton, R. B., Sands, M. *The Feynman Lectures on Physics*, Vol. ll. Reading, MA: Addison-Wesley, 1963.

23. Frenkel, J., Doerman, J. Spontaneous and induced magnetisation in ferromagnetic bodies. *Nature* 1930, 126, 274–5.

24. Han, D. H., Wang, J. P., Luo, H. L. Crystallite size effect on saturation magnetization of fine ferrimagnetic particles. *J Magn Magn Mater* 1994, 136, 176–82.

25. Bate G. Particulate recording materials. *Proceedings of the IEEE* 1986, 74, 1513–25.

26. Serna, C. J., Morales, M. P. Maghemite (γ-Fe_2O_3): A versatile magnetic colloidal material. In *Surface and Colloid Science*, (Eds.) E. Matijevic, M. Borkovec, Kluwer Academic/Plenum Publishers, 2004, pp. 17, Chapter 2.

27. Marchal, G., Van Hecke, P., Demaerel, P., Decrop, E., Kennis, C., Baert, A. L., van der Schueren, E. Detection of liver metastases with superparamagnetic iron oxide in 15 patients: Results of MR imaging at 1.5 T. *AJR Am J Roentgenol* 1989, 152, 771–5.

28. Weissleder, R., Stark, D. D., Engelstad, B. L., Bacon, B. R., Compton, C. C., White, D. L., Jacobs, P., Lewis, J. Superparamagnetic iron oxide: Pharmacokinetics and toxicity. *AJR Am J Roentgenol* 1989, 152, 167–73.

29. Carenza, E., Barcelo, V., Morancho, A., Levander, L., Boada, C., Laromaine, A., Roig, A., Montaner, J., Rosell, A. *In vitro* angiogenic performance and in vivo brain targeting of magnetized endothelial progenitor cells for neurorepair therapies. *Nanomedicine* 2014, 10, 225–34.

30. Cheng, K., Malliaras, K., Li, T. S., Sun, B., Houde, C., Galang, G., Smith, J., Matsushita, N., Marban, E. Magnetic enhancement of cell retention, engraftment, and functional benefit after intracoronary delivery of cardiac-derived stem cells in a rat model of ischemia/reperfusion. *Cell Transplant* 2012, 21, 1121–35.

31. Kang, H. J., Kim, J. Y., Lee, H. J., Kim, K. H., Kim, T. Y., Lee, C. S., Lee, H. C., Park, T. H., Kim, H. S., Park, Y. B. Magnetic bionanoparticle enhances homing of endothelial progenitor cells in mouse hindlimb ischemia. *Korean Cir J* 2012, 42, 390–6.

32. Kyrtatos, P. G., Lehtolainen, P., Junemann-Ramirez, M., Garcia-Prieto, A., Price, A. N., Martin, J. F., Gadian, D. G., Pankhurst, Q. A., Lythgoe, M. F. Magnetic tagging increases delivery of circulating progenitors in vascular injury. *JACC Cardiovasc Interv* 2009, 2, 794–802.

33. Luciani, A., Wilhelm, C., Bruneval, P., Cunin, P., Autret, G., Rahmouni, A., Clement, O., Gazeau, F. Magnetic targeting of iron-oxide-labeled fluorescent hepatoma cells to the liver. *Eur Radiol* 2009, 19, 1087–96.
34. Song, M., Kim, Y. J., Kim, Y. H., Roh, J., Kim, S. U., Yoon, B. W. Using a neodymium magnet to target delivery of ferumoxide-labeled human neural stem cells in a rat model of focal cerebral ischemia. *Hum Gene Ther* 2010, 21, 603–10.
35. Yanai, A., Hafeli, U. O., Metcalfe, A. L., Soema, P., Addo, L., Gregory-Evans, C. Y., Po, K., Shan, X., Moritz, O. L., Gregory-Evans, K. Focused magnetic stem cell targeting to the retina using superparamagnetic iron oxide nanoparticles. *Cell Transplant* 2012, 21, 1137–48.
36. Chaudeurge, A., Wilhelm, C., Chen-Tournoux, A., Farahmand, P., Bellamy, V., Autret, G., Menager, C., Hagege, A., Larghero, J., Gazeau, F., Clement, O., Menasche, P. Can magnetic targeting of magnetically labeled circulating cells optimize intramyocardial cell retention? *Cell Transplant* 2012, 21, 679–91.
37. Hori, J., Deie, M., Kobayashi, T., Yasunaga, Y., Kawamata, S., Ochi, M. Articular cartilage repair using an intra-articular magnet and synovium-derived cells. *J Orthop Res* 2011, 29, 531–8.
38. Nishida, K., Tanaka, N., Nakanishi, K., Kamei, N., Hamasaki, T., Yanada, S., Mochizuki, Y., Ochi, M. Magnetic targeting of bone marrow stromal cells into spinal cord: Through cerebrospinal fluid. *Neuroreport* 2006, 17, 1269–72.
39. Sasaki, H., Tanaka, N., Nakanishi, K., Nishida, K., Hamasaki, T., Yamada, K., Ochi, M. Therapeutic effects with magnetic targeting of bone marrow stromal cells in a rat spinal cord injury model. *Spine* 2011, 36, 933–8.
40. Fujioka, Y., Tanaka, N., Nakanishi, K., Kamei, N., Nakamae, T., Izumi, B., Ohta, R., Ochi, M. Magnetic field-based delivery of human CD133(+) cells promotes functional recovery after rat spinal cord injury. *Spine* 2012, 37, E768–77.
41. Kobayashi, T., Ochi, M., Yanada, S., Ishikawa, M., Adachi, N., Deie, M., Arihiro, K. A novel cell delivery system using magnetically labeled mesenchymal stem cells and an external magnetic device for clinical cartilage repair. *Arthroscopy* 2008, 24, 69–76.
42. Kodama, A., Kamei, N., Kamei, G., Kongcharoensombat, W., Ohkawa, S., Nakabayashi, A., Ochi, M. In vivo bioluminescence imaging of transplanted bone marrow mesenchymal stromal cells using a magnetic delivery system in a rat fracture model. *J Bone Joint Surg Br* 2012, 94, 998–1006.
43. Ohkawa, S., Kamei, N., Kamei, G., Shi, M., Adachi, N., Deie, M., Ochi, M. Magnetic targeting of human peripheral blood CD133+ cells for skeletal muscle regeneration. *Tissue Eng Part C Methods* 2013, 19, 631–41.
44. Riegler, J., Liew, A., Hynes, S. O., Ortega, D., O'Brien, T., Day, R. M., Richards, T., Sharif, F., Pankhurst, Q. A., Lythgoe, M. F. Superparamagnetic iron oxide nanoparticle targeting of MSCs in vascular injury. *Biomaterials* 2013, 34, 1987–94.
45. Pislaru, S. V., Harbuzariu, A., Gulati, R., Witt, T., Sandhu, N. P., Simari, R. D., Sandhu, G. S. Magnetically targeted endothelial cell localization in stented vessels. *J Am Coll Cardiol* 2006, 48, 1839–45.
46. Polyak, B., Fishbein, I., Chorny, M., Alferiev, I., Williams, D., Yellen, B., Friedman, G., Levy, R. J. High field gradient targeting of magnetic nanoparticle-loaded endothelial cells to the surfaces of steel stents. *Proc Natl Acad Sci U S A* 2008, 105, 698–703.

47. Alexiou, C., Diehl, D., Henninger, P., Iro, H., Rockelein, R., Schmidt, W., Weber, H. A High field gradient magnet for magnetic drug targeting. *IEEE Trans Appl Supercond* 2006, 16, 1527–30.

48. Halbach, K. Design of permanent multipole magnets with oriented rare earth cobalt material. *Nucl Instrum Methods* 1980, 169, 1–10.

49. Johnson, B., Toland, B., Chokshi, R., Mochalin, V., Koutzaki, S., Polyak, B. Magnetically responsive paclitaxel-loaded biodegradable nanoparticles for treatment of vascular disease: Preparation, characterization and *in vitro* evaluation of anti-proliferative potential. *Curr Drug Deliv* 2010, 7, 263–73.

50. Kawaguchi, T., Hasegawa, M. Structure of dextran-magnetite complex: Relation between conformation of dextran chains covering core and its molecular weight. *J Mater Sci Mater Med* 2000, 11, 31–5.

51. MacDonald, C., Barbee, K., Polyak, B. Force dependent internalization of magnetic nanoparticles results in highly loaded endothelial cells for use as potential therapy delivery vectors. *Pharm Res* 2012, 29, 1270–81.

52. Orynbayeva, Z., Sensenig, R., Polyak, B. Metabolic and structural integrity of magnetic nanoparticle-loaded primary endothelial cells for targeted cell therapy. *Nanomedicine (Lond)* 2015, 10, 1555–68.

53. Zohra, F. T., Medved, M., Lazareva, N., Polyak, B. Functional behavior and gene expression of magnetic nanoparticle-loaded primary endothelial cells for targeting vascular stents. *Nanomedicine (Lond)* 2015, 10, 1391–406.

54. McJunkin, F. A. The phagocytic activity of the vascular endothelium of granulation tissue. *Am J Pathol* 1928, 4, 587–92.1.

55. Arbab, A. S., Yocum, G. T., Kalish, H., Jordan, E. K., Anderson, S. A., Khakoo, A. Y., Read, E. J., Frank, J. A. Efficient magnetic cell labeling with protamine sulfate complexed to ferumoxides for cellular MRI. *Blood* 2004, 104, 1217–23.

56. Gamarra, L. F., Pavon, L. F., Marti, L. C., Pontuschka, W. M., Mamani, J. B., Carneiro, S. M., Camargo-Mathias, M. I., Moreira-Filho, C. A., Amaro, E., Jr. *In vitro* study of CD133 human stem cells labeled with superparamagnetic iron oxide nanoparticles. *Nanomedicine* 2008, 4, 330–9.

57. Landazuri, N., Tong, S., Suo, J., Joseph, G., Weiss, D., Sutcliffe, D. J., Giddens, D. P., Bao, G., Taylor, W. R. Magnetic targeting of human mesenchymal stem cells with internalized superparamagnetic iron oxide nanoparticles. *Small* 2013, 9, 4017–26.

58. Politi, L. S., Bacigaluppi, M., Brambilla, E., Cadioli, M., Falini, A., Comi, G., Scotti, G., Martino, G., Pluchino, S. Magnetic-resonance-based tracking and quantification of intravenously injected neural stem cell accumulation in the brains of mice with experimental multiple sclerosis. *Stem Cells* 2007, 25, 2583–92.

59. Cohen, M. E., Muja, N., Fainstein, N., Bulte, J. W., Ben-Hur, T. Conserved fate and function of ferumoxides-labeled neural precursor cells *in vitro* and in vivo. *J Neurosci Res* 2010, 88, 936–44.

60. Taylor, A., Wilson, K. M., Murray, P., Fernig, D. G., Levy, R. Long-term tracking of cells using inorganic nanoparticles as contrast agents: Are we there yet? *Chem Soc Rev* 2012, 41, 2707–17.

61. Cromer Berman, S. M., Walczak, P., Bulte, J. W. Tracking stem cells using magnetic nanoparticles. *Wiley Interdiscip Rev Nanomed Nanobiotechnol* 2011, 3, 343–55.

62. Caplan, A. I., Dennis, J. E. Mesenchymal stem cells as trophic mediators. *J Cell Biochem* 2006, 98, 1076–84.

63. Schraufstatter, I. U., Discipio, R. G., Khaldoyanidi, S. Mesenchymal stem cells and their microenvironment. *Front Biosci (Landmark Ed)* 2011, 16, 2271–88.

64. Adams, C. F., Rai, A., Sneddon, G., Yiu, H. H., Polyak, B., Chari, D. M. Increasing magnetite contents of polymeric magnetic particles dramatically improves labeling of neural stem cell transplant populations. *Nanomedicine* 2015, 11, 19–29.

65. Vanecek, V., Zablotskii, V., Forostyak, S., Ruzicka, J., Herynek, V., Babic, M., Jendelova, P., Kubinova, S., Dejneka, A., Sykova, E. Highly efficient magnetic targeting of mesenchymal stem cells in spinal cord injury. *Int J Nanomed* 2012, 7, 3719–30.

66. Adamo, R. F., Fishbein, I., Zhang, K., Wen, J., Levy, R. J., Alferiev, I. S., Chorny, M. Magnetically enhanced cell delivery for accelerating recovery of the endothelium in injured arteries. *J Control Release* 2016, 222, 169–75.

67. Tefft, B. J., Uthamaraj, S., Harburn, J. J., Hlinomaz, O., Lerman, A., Dragomir-Daescu, D., Sandhu, G. S. Magnetizable stent-grafts enable endothelial cell capture. *J Magn Magn Mater* 2017, 427, 100–104.

68. Polyak, B., Medved, M., Lazareva, N., Steele, L., Patel, T., Rai, A., Rotenberg, M. Y., Wasko, K., Kohut, A. R., Sensenig, R., Friedman, G. Magnetic nanoparticle-mediated targeting of cell therapy reduces in-stent stenosis in injured arteries. *ACS Nano* 2016, 10 (10), 9559–69.

69. Vosen, S., Rieck, S., Heidsieck, A., Mykhaylyk, O., Zimmermann, K., Bloch, W., Eberbeck, D., Plank, C., Gleich, B., Pfeifer, A., Fleischmann, B. K., Wenzel, D. Vascular repair by circumferential cell therapy using magnetic nanoparticles and tailored magnets. *ACS Nano* 2016, 10 (1), 369–76.

70. Arbab, A. S., Jordan, E. K., Wilson, L. B., Yocum, G. T., Lewis, B. K., Frank, J. A. In vivo trafficking and targeted delivery of magnetically labeled stem cells. *Hum Gene Ther* 2004, 15, 351–60.

71. Shoji, M., Sata, M., Fukuda, D., Tanaka, K., Sato, T., Iso, Y., Shibata, M., Suzuki, H., Koba, S., Geshi, E., Katagiri, T. Temporal and spatial characterization of cellular constituents during neointimal hyperplasia after vascular injury: Potential contribution of bone-marrow-derived progenitors to arterial remodeling. *Cardiovasc Pathol* 2004, 13, 306–12.

72. Smith, S. H., Geer, J. C. Morphology of saphenous vein-coronary artery bypass grafts: Seven to 116 months after surgery. *Arch Pathol Lab Med* 1983, 107, 13–8.

73. Roiron, C., Sanchez, P., Bouzamondo, A., Lechat, P., Montalescot, G. Drug eluting stents: An updated meta-analysis of randomised controlled trials. *Heart* 2006, 92, 641–9.

74. Muradin, G. S., Bosch, J. L., Stijnen, T., Hunink, M. G. Balloon dilation and stent implantation for treatment of femoropopliteal arterial disease: Meta-analysis. *Radiology* 2001, 221, 137–45.

75. Aoki, J., Serruys, P. W., van Beusekom, H., Ong, A. T., McFadden, E. P., Sianos, G., van der Giessen, W. J., Regar, E., de Feyter, P. J., Davis, H. R., Rowland, S., Kutryk, M. J. Endothelial progenitor cell capture by stents coated with antibody against CD34: The HEALING-FIM (Healthy Endothelial Accelerated Lining Inhibits Neointimal Growth-First In Man) registry. *J Am Coll Cardiol* 2005, 45, 1574–9.

76. Walter, D. H., Cejna, M., Diaz-Sandoval, L., Willis, S., Kirkwood, L., Stratford, P. W., Tietz, A. B., Kirchmair, R., Silver, M., Curry, C., Wecker, A., Yoon, Y. S., Heidenreich, R., Hanley, A., Kearney, M., Tio, F. O., Kuenzler, P., Isner, J. M., Losordo, D. W. Local gene transfer of phVEGF-2 plasmid by gene-eluting stents: An alternative strategy for inhibition of restenosis. *Circulation* 2004, 110, 36–45.

77. Landmesser, U., Hornig, B., Drexler, H. Endothelial function: A critical determinant in atherosclerosis? *Circulation* 2004, 109, II27–33.
78. Patti, G., Pasceri, V., Melfi, R., Goffredo, C., Chello, M., D'Ambrosio, A., Montesanti, R., Di Sciascio, G. Impaired flow-mediated dilation and risk of restenosis in patients undergoing coronary stent implantation. *Circulation* 2005, 111, 70–5.
79. Werner, N., Junk, S., Laufs, U., Link, A., Walenta, K., Bohm, M., Nickenig, G. Intravenous transfusion of endothelial progenitor cells reduces neointima formation after vascular injury. *Circ Res* 2003, 93, e17–24.
80. Patel, S. D., Waltham, M., Wadoodi, A., Burnand, K. G., Smith, A. The role of endothelial cells and their progenitors in intimal hyperplasia. *Ther Adv Cardiovasc Dis* 2010, 4, 129–41.
81. Casscells, W. Migration of smooth muscle and endothelial cells. Critical events in restenosis. *Circulation* 1992, 86, 723–9.
82. Reidy, M. A., Clowes, A. W., Schwartz, S. M. Endothelial regeneration. V. Inhibition of endothelial regrowth in arteries of rat and rabbit. *Lab Invest* 1983, 49, 569–75.
83. Schwartz, R. S., Holmes, D. R., Jr., Topol, E. J. The restenosis paradigm revisited: An alternative proposal for cellular mechanisms. *J Am Coll Cardiol* 1992, 20, 1284–93.
84. Lindner, V., Reidy, M. A., Fingerle, J. Regrowth of arterial endothelium. Denudation with minimal trauma leads to complete endothelial cell regrowth. *Lab Invest* 1989, 61, 556–63.
85. Clowes, A. W., Reidy, M. A., Clowes, M. M. Kinetics of cellular proliferation after arterial injury. I. Smooth muscle growth in the absence of endothelium. *Lab Invest* 1983, 49, 327–33.
86. Asahara, T., Chen, D., Tsurumi, Y., Kearney, M., Rossow, S., Passeri, J., Symes, J. F., Isner, J. M. Accelerated restitution of endothelial integrity and endothelium-dependent function after phVEGF165 gene transfer. *Circulation* 1996, 94, 3291–302.
87. Hedman, M., Hartikainen, J., Syvanne, M., Stjernvall, J., Hedman, A., Kivela, A., Vanninen, E., Mussalo, H., Kauppila, E., Simula, S., Narvanen, O., Rantala, A., Peuhkurinen, K., Nieminen, M. S., Laakso, M., Yla-Herttuala, S. Safety and feasibility of catheter-based local intracoronary vascular endothelial growth factor gene transfer in the prevention of postangioplasty and in-stent restenosis and in the treatment of chronic myocardial ischemia: Phase II results of the Kuopio Angiogenesis Trial (KAT). *Circulation* 2003, 107, 2677–83.
88. Losordo, D. W., Isner, J. M., Diaz-Sandoval, L. J. Endothelial recovery: The next target in restenosis prevention. *Circulation* 2003, 107, 2635–7.
89. Pislaru, S. V., Harbuzariu, A., Agarwal, G., Witt, T., Gulati, R., Sandhu, N. P., Mueske, C., Kalra, M., Simari, R. D., Sandhu, G. S. Magnetic forces enable rapid endothelialization of synthetic vascular grafts. *Circulation* 2006, 114, I314–8.
90. McBain, S. C., Griesenbach, U., Xenariou, S., Keramane, A., Batich, C. D., Alton, E. W., Dobson, J. Magnetic nanoparticles as gene delivery agents: Enhanced transfection in the presence of oscillating magnet arrays. *Nanotechnology* 2008, 19, 405102.
91. Pickard, M., Chari, D. Enhancement of magnetic nanoparticle-mediated gene transfer to astrocytes by 'magnetofection': Effects of static and oscillating fields. *Nanomedicine (Lond)* 2010, 5, 217–32.

7

Magnetic Capture and Actuation of Thermosensitive Drug Carriers Using Iron Oxide Nanoparticles

Mary Kathryn Sewell-Loftin, Mary L. Hampel, Amy E. Frees,
Lauren M. Blue, Natalie Lapp, Minghua Zhang, Jaimee M. Robertson,
Rhythm R. Shah, and Christopher S. Brazel

CONTENTS

7.1 Introduction... 148
    7.1.1 Targeting Mechanisms 148
    7.1.2 Triggering Mechanisms: Thermosensitive Release.................. 151
7.2 Materials and Methods.. 153
    7.2.1 Materials for Magnetic Capture.. 153
    7.2.2 Experimental Magnetic Capture in Flow 153
    7.2.3 Opsonization of MNPs.. 154
    7.2.4 MNP Synthesis 155
    7.2.5 Hydrogel Synthesis 156
    7.2.6 MNP Retention in Hydrogels.. 156
    7.2.7 Equilibrium Swelling.. 156
    7.2.8 Drug Loading 157
    7.2.9 Constant Temperature and Pulsatile
          Drug Release 157
    7.2.10 Magnetic Heating... 158
7.3 Results and Discussion 158
    7.3.1 Magnetic Capture of Aqueous-Dispersed NPs 158
    7.3.2 Effect of Tubing Diameter and Viscosity Modifiers on
          Magnetic Capture.. 159
    7.3.3 Effect of Protein Opsonization.. 160
    7.3.4 Magnetic Hydrogels: MNP Synthesis and
          Dispersion 162
    7.3.5 Gel Synthesis... 162
    7.3.6 MNP Retention in Gels... 163
    7.3.7 Equilibrium Swelling.. 163

        7.3.8   Drug Release 165
        7.3.9   Magnetic Heating 167
7.4   Conclusions 169
Acknowledgments 170
References 171

7.1 Introduction

A number of strategies to treat disease at the cellular level involve targeting drug release systems within the body. For example, chemotherapy is a common cancer treatment, but it involves administering potent drugs that cause side effects so debilitating that the patient must sometimes discontinue the therapy before the cancer is eradicated.[1] Many of these side effects are exacerbated by the free administration of these drugs, allowing the drugs to kill healthy cells as well as tumor cells. The development of targeted and triggered delivery systems medication allows physicians to gain control over the location and time of drug delivery. The actuation of targeting and release using magnetic fields is the focus of this paper.

Magnetic nanoparticles (MNPs) are receiving special attention in cancer research because of their dual functions in diagnostic imaging[2,3] and therapy.[4,5] On the therapeutic side, MNPs can be employed to assist with targeting and drug localization through magnetic capture using static magnets to accumulate MNPs (or nanogels loaded with MNPs) in a specific region of the body.[6] Magnetic fluid hyperthermia (MFH), by which magnetic fields are used to induce local heating in tissue, has progressed to clinical usage in some instances.[7] Additionally, MNPs can serve as heat sources as part of a magnetothermal triggering mechanism to initiate drug release from thermosensitive hydrogels after the MNPs are localized.[5] This chapter briefly reviews targeting and triggered release strategies, assesses the use of MNPs in magnetic localization and heating-activated release, and presents experimental results for MNP systems used for capture and drug release.

7.1.1 Targeting Mechanisms

There are numerous ways for selective targeting within the body; see for instance an excellent review of targeting modalities by Fahmy et al.[8] Antibodies attached to drug carriers can specifically target certain types of cells and have been commercialized for a wide range of therapies (Table 7.1). They are typically administered intravenously and travel through the bloodstream to reach the site of action. The antibodies attach to specific receptors, which are transmembrane proteins that extend outside cell membranes and with expression dependent on cell differentiation. This is particularly useful for cancer

TABLE 7.1

Selected Ligand-Targeted Nanoparticle Systems for Delivery of Therapeutics

Ligand	Targeted Cells/ Tissue	Carrier System	Drug
Nucleic Acids			
Aptamers	Prostrate epithelial cells	Poly(lactic acid)	
Aptamers[14]	Macula/eye	Poly(ethylene glycol) (PEG)	Pegaptanib sodium
Proteins			
Integrin	Melanoma cells	Liposomes	Raf genes
RGD peptides	Tumor vasculature	Poly(ethylene imine)	siRNA
Fibrinogen	Tumor vasculature	Albumin	Radioisotopes
Von Willebrand Factor	Pancreatic cancers	Viral particles	Cyclin gene
Phage peptides[15]	Cancers	PEGylated liposomes	Doxorubicin
Lipids			
MP Lipid A	Dendritic cells	Poly(lactic acid-co-glycolic acid)	
Carbohydrates			
Galactose	Hepatocytes	Poly(lactic acid)	Retinoic acid
Hyaluronic Acid	CD44+ melanoma cells	Liposomes	Doxorubicin
Peptidomimetics	Brain cells	PEG/Poly(lactic acid-co-glycolic acid)	Numerous
Antibodies to			
Human epidermal growth factor receptor 2 (HER2) receptor	HER2 breast cancer cells	Gelatin/ hydroxyapatite; liposomes	Doxorubicin
CD19	B cell lymphoma	Liposomes	Doxorubicin
P. falciparum infected parasitic red blood cells[16]	Malarial red blood cells	Immunoliposomes	Chloroquine
Vitamins			
Folate		Liposomes	Docorubicin
Blood Proteins			
Albumin		Albumin/drug conjugate	Paclitaxel

Source: Modified and supplemented from T. M. Fahmy et al., *Mater. Today*, 8, 18–26, 2005.

therapies, where various types of cancers have been found to express large quantities of receptors varying in specificity, such as integrin or folic acid receptors.[9,10] Such receptor–antibody binding and the cell-uptake mechanisms have been reviewed extensively,[8] and many are used in targeted therapies that are now approved by the Food and Drug Administration (FDA).[11] For cancer therapies in particular, the enhanced permeation and retention (EPR) effect has also proven to be an effective way to localize chemotherapy medicine to a growing tumor.[12] Because growing tumors are in the process of angiogenesis, a leaky vasculature leads to blood flowing into the tumor and not having a pathway to leave. For these reasons, even passive targeting is helpful in localizing cancer therapeutics, allowing nanoparticles (NPs) to pass from the blood vessel into the cancerous tissue. Thus, intravenously administered medication will preferentially end up in a tumor, particularly if it is of an optimal size and coated with poly(ethylene glycol) (PEG) to avoid opsonization and removal through the filtration organs (liver and kidney).[13]

Magnetic localization of drug carriers has been investigated over the past few decades, with varying levels of success. A number of experimental and theoretical studies have investigated the use of static magnets to capture magnetic microparticles and NPs flowing in fluids ranging from water to blood.[6,17,18] Animal studies have proven that MNPs indeed enhance accumulation in tumors; experiments have shown that using MNPs can increase targeting by several fold compared to the administration of a free drug.[19]

MNPs are captured when the magnetic forces overcome the momentum of the particles flowing in the bloodstream. NP aggregation tends to improve capture, although this can be problematic if the aggregate size causes blockage or activates the immune system.[18] Doshi et al.[20] found that the MNP shape affects the efficiency of entrapment. With particle volume and surface area constant, their study showed that elongated particles showed better adhesion to the capillary walls than spherical MNPs. Elongated disc-shaped particles not only accumulate more efficiently, but are also less likely to accumulate in the liver or other organs.[21]

There are many challenges to successful magnetic localization. For one, the strength of a static magnetic field diminishes sharply with distance, so excessively strong magnets would be needed to reach significant subcutaneous penetration depths. Also, since the highest magnetic field strength is near the skin, efficient targeting to deep tissues will result in the simultaneous accumulation of magnetic particles near the skin. Third, permanent magnets placed *in vivo* tend to aggregate, creating potential problems for cardiovascular delivery, as blockage or thrombotic events may occur. Finally, magnetic localization requires knowledge of the location of the targeted tissue; while this would likely be true for developed tumors or diseases located on specific organs or tissues, magnetic targeting does not offer the cellular specificity of antibodies; thus, magnetic localization will not be useful for targeting metastatic cancers. As a way to optimize accumulation of MNPs, particularly in deeper tissues, magnetic implants have been proposed which can collect MNPs near a target organ or tissue.[22] This has shown some success; however, the requirement to place magnetic surfaces *in vivo* negates the inherent advantages of magnetic localization (namely treatment that uses only an injection and is minimally invasive). If some of these obstacles can be overcome, magnetic localization could be an important tool in optimizing the therapeutic efficacy of many drugs.

Here, we investigate the capture of MNPs using static magnetic fields in flow environments to simulate blood vessels. Capture efficiency was determined as a function of vessel diameter, flow rate, MNP dispersion, presence of blood proteins, and magnetic field strength.

7.1.2 Triggering Mechanisms: Thermosensitive Release

Optimal delivery of medication includes reaching the location of the disease with the correct dosage. While magnetic capture is one approach for localization, magnetic actuation can also be used to trigger drug release. For effective drug therapy using targeted systems, the majority of drug must remain sequestered in the delivery vehicle to minimize side effects (particularly with a potent drug) until it reaches the desired site of action. After this time, a trigger for drug release is required. While some systems have approached this problem by designing materials that release drug when broken down upon entering cells, an external triggering mechanism is a viable alternative. The mechanism used for triggering release can be subtle changes in physiological conditions such as pH or ionic strength, but the use of MNPs imparts special significance to thermally sensitive polymers as numerous groups have developed iron oxide-based NPs that heat efficiently when exposed to radio-frequency magnetic fields.[23,24] Out of several thermally sensitive polymers, those finding the greatest potential for use in medical applications are those with phase separation temperatures near 37°C. This includes copolymers of poly(N-isopropylacrylamide)[25] and triblock copolymers of poly(ethylene glycol-b-propylene glycol-b-ethylene glycol), sold commercially as Pluronic®.[26] The mechanism by which these polymer systems works is based on thermally induced phase transition, which occurs when the polymer is heated above the lower critical solution temperature (LCST).

The ability to externally control the timed release of a medication remains a major challenge for the design and development of new materials. Some of the main approaches to trigger drug delivery after administration to a patient include using photodynamic or photothermally responsive materials,[27,28] ultrasonically rupturable structures,[20–30] electronic devices,[31,32] and magnetothermally triggered release.[33] These methods for external triggered release have been reviewed recently with a focus on magnetic materials.[5] For example, a novel platform for cancer treatment has been proposed that localizes chemotherapy by targeting tumors using folate receptors.[34] After localization, either a near-IR light source (using gold nanoshells) or an AC magnetic field (using MNPs) can be used to locally heat NPs imbedded in the release device, which in turn causes a phase separation in the material to trigger release of the chemotherapy agent. One advantage of such a system is that the localized heating that triggers drug release can also result in localized hyperthermia, which has also been shown to effectively treat cancers.[35,36] Combined therapies have been shown to greatly improve patient response to cancer treatment,[37] and localized triggering should further increase the efficacy of cancer treatment, while decreasing negative side effects, which in turn should improve the overall quality of life for patients undergoing treatment.

In addition to magnetothermally triggered release, hyperthermia on its own (or in combination with chemotherapy or radiation) has proven to be an effective treatment for cancers specifically defined as a heating tissue to temperatures between 42°C and 45°C for a period in the order of minutes to hours to preferentially kill cancer cells.[4,38,39] Hyperthermia leads to cell death that occurs almost exclusively in cancer cells because tumor tissues do not have the fully formed blood network needed for efficient cooling, as healthy tissue does.[1] Although hyperthermia therapy for cancer has received less attention compared to chemotherapy, radiation and surgery, advances in localized heating using gold[40] and magnetic[7] NPs has resulted in several human clinical trials in recent years. Iron-containing nanoparticles (MNPs) can be used in MFH, as they are effectively heated using high frequency magnetic fields.[41,42] The mechanism of heating is based on Néel relaxation inside nano-sized magnetic crystals, as detailed by Rosensweig.[43] The magnitude of heating is dependent on magnetic particle composition and size as well as the properties of the applied AC magnetic field.[44]

Environmentally responsive polymers have been studied for decades as a way to increase the efficacy of drug delivery, decrease administration frequency, and decrease the side effects of potent medications.[45–48] These materials have thermodynamic properties that cause them to respond to a change in temperature, pH, ion concentration, or other variances in solution properties.[49] The ability to trigger heating using a magnetic field brings forth a novel way to thermally activate release using LCST polymers. For heat-triggered release, a key component is the polymer structure that houses the MNPs and carries the drug. On/off modulated drug release has been extensively investigated over the past two decades, with copolymers of N-isopropylacrylamide (NIPAAm) showing the most promise for biological applications due to their sharp phase transition at temperatures near physiological conditions.[50,51] Crosslinked PNIPAAm hydrogels display an LCST in aqueous solutions, swelling when hydrophilic at low temperatures and collapsing (phase separating) when heated above the LCST, where hydrophobic moieties become the dominant molecular features.[52] Drugs imbedded in PNIPAAm-based hydrogels, therefore, have been shown to squeeze out drug in a rapid burst when heated. By selecting appropriate comonomers to polymerize with PNIPAAm, the LCST can be changed from approximately 32°C (for PNIPAAm homopolymer) to higher (with hydrophilic comonomers) or lower (with hydrophobic comonomers) temperatures.[25] For heat-activated release *in vivo*, the transition temperature must be adjusted to slightly above physiological conditions. For magnetically heated systems, the desired transition temperature (or LCST) should be in the range of temperatures used for hyperthermia, or around 42°C. Here, poly(NIPAAm-co-acrylamide) is a suitable choice, as polyacrylamide is more hydrophilic, and the LCST can be adjusted (or tuned) to values higher than 32°C by altering the ratio and configuration of acrylamide:NIPAAm monomers in the hydrogel structure.[45] Here, we investigate composite materials consisting of thermally responsive hydrogels with dispersed MNPs to determine their feasibility and design requirements for use in a magnetothermally triggered drug delivery system.

Targeting MNPs in the human body (whether by passive targeting, magnetic capture, or active targeting of receptors) has the potential to deliver two modes of therapy at the same location: hyperthermia caused by AC magnetic field heating and triggered release of chemotherapy drugs or other medications that results from the localized heating. This chapter reports on the influence of flow, dispersion, and magnetic field on magnetic capture, as well as the design and evaluation of hydrogels that contain dispersed MNPs for triggered release of an anticancer agent upon a heating.

7.2 Materials and Methods

7.2.1 Materials for Magnetic Capture

For magnetic capture, maghemite nanopowder (Aldrich, St. Louis, MO) was used with different coatings, including dopamine, sulfosuccinic acid (both from Aldrich), and PEG. The nanopowder MNPs had average diameters of 50 nm (as determined by transmission electron microscopy). Viscosity modifiers included PEG with molecular weight of 8000 (Union Carbide), as well as several blood proteins, including albumin and γ-globulins (both from bovine sources through Sigma-Aldrich). Fetal bovine serum (Gibco, Invitrogen) was used to test magnetic capture in simulated blood. Static neodymium magnets of various sizes and field strengths were purchased from K+J Magnetics (Pipersville, PA).

7.2.2 Experimental Magnetic Capture in Flow

Magnetic capture was investigated using the set-up shown in Figure 7.1. Aqueous MNP solutions were made by mixing MNPs and surfactants at a 1:4 weight ratio (to achieve MNP concentrations from 0.02 to 0.05 g/L) in water, which were subjected to sonication for 30–60 minutes to achieve dispersions. In some instances, PEG or proteins were added prior to sonication to mimic the viscosity and/or the components of blood sera. Tygon tubing was used to circulate MNP solutions from a continuously mixing source vessel, passing through a section of tubing with constant inner diameters ranging from 0.5 to 3.0 mm, where a magnetic field could be applied. The magnetic field was measured as a function of distance from the surface of the magnet using a Gaussmeter (Model GM-1-ST, AlphaLab, Inc.) and is reported as the magnetic field intensity at the center line of the tubing. After passing through the magnetic capture section, the magnetic fluid passes through a flow-through cuvette inside a UV-Vis spectrophotometer (Shimadzu, UV-2401PC) where the absorbance of the magnetic fluid was measured before the fluid returned to the source vessel. Known NP concentrations were used to determine a

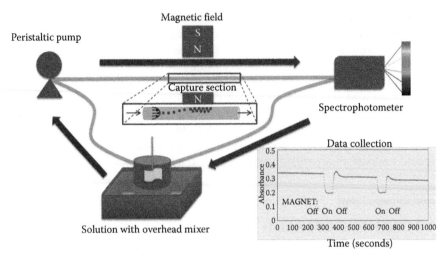

FIGURE 7.1
Diagram of experimental set-up for magnetic capture.

calibration curve according to Beer's law, using 500 nm to monitor absorbance (or 700 nm in the case of fetal bovine serum [FBS], to avoid overlap with absorbance due to proteins in the serum).

After a consistent flow was established throughout the system, and a baseline absorbance measured, a static magnet was placed outside the tubing for one minute, yielding a known magnetic field at the centerline of flow, then released. During this time, MNPs were collected, reducing the measured absorbance downstream from the testing region (Figure 7.1). The drop in absorbance was integrated to determine the mass of NPs captured. After releasing the magnetic field, the system was allowed to reach a stable baseline absorbance before subsequent applications of magnetic field. In all capture experiments, three replicates were used, with the average and standard deviation reported.

Experimental parameters investigated for magnetic capture included: linear flow velocity, magnetic field strength, inside tubing diameter (vessel diameter), presence of viscosity modifiers PEG, or blood proteins, and dispersion in FBS.

7.2.3 Opsonization of MNPs

The attachment of blood proteins to MNPs was monitored using dynamic light scattering (DLS, Malvern Instruments, ZEN3600). Maghemite nanopowder was first dispersed in an aqueous solution, then exposed to a saline solution containing 0.30 g/L albumin and 0.15 g/L α- and γ-globulins, with number-average particle size reported as a function of time after protein exposure. DLS measurements were taken of the protein solution, the MNP solution, and the solutions after an elapsed time. A commercially available MNP dispersion, fluidMAG-D® (chemicell, Berlin, Germany), which consists of magnetite NPs coated with dextran, was used for comparison.

7.2.4 MNP Synthesis

Several different MNP compositions were tested for incorporation into polymer gels. In addition to the maghemite nanopowder discussed above, iron platinum and cobalt ferrite MNPs were synthesized in-house and combined with LCST polymer hydrogels.

Iron platinum NPs were prepared by a thermal decomposition method.[53,54] Briefly, equal moles of iron (II) chloride (Aldrich, St. Louis, MO) and platinum (II) acetylacetonate (Aldrich) were dissolved in diphenyl ether (Acros, Fair Lawn, NJ) and heated to 100°C under nitrogen and mixed with a small amount of oleic acid (Fisher, Fair Lawn, NJ) and oleyl amine (Aldrich). Upon addition of a reducing agent, the mixture was further heated to the boiling point of phenyl ether and refluxed at 255°C for 1.5 hours. The particles were recovered by successive precipitation and redispersion in ethanol and hexane, respectively. A ligand-exchange reaction was used to convert the particles dispersed in hexane to a stable aqueous dispersion. The oleic acid and oleyl amine residues were replaced with mercaptoundecanoic acid (Aldrich) by forming a two-phase aqueous/organic mixture, and the ligand-exchanged particles were dispersed in the aqueous phase. $Fe_{33}Pt_{67}$ particles (referred to as FePt throughout the chapter) with 4–8 nm diameters (as confirmed by transmission electron microscopy) were used.[54] The aqueously dispersed FePt MNPs were thoroughly characterized by x-ray diffractometry, alternating gradient magnetometry, infrared spectroscopy, and electron microscopy, as detailed by Bagaria et al.[54]

Cobalt ferrite NPs were prepared by a similar technique as reported by Sun et al.[55] Briefly, iron (III) acetylacetonate was dissolved in benzyl ether (Aldrich) and mixed with surfactants oleic acid and oleylamine, followed by the addition of cobalt (II) acetylacetonate (Aldrich) and 1,2-hexanediol (Aldrich). The reaction was carried out under nitrogen in a round bottom flask with a condenser fitted to allow reflux at 285°C after reacting first at 100°C and 200°C for fixed periods of time. Ethanol was added to the solution, and the NPs were recovered after centrifugation. Similar steps to the treatment of FePt were then taken to exchange the ligands on the MNP surfaces from oleic acid/oleyl amine to a water-dispersible surfactant—meso-2,3-dimercaptosuccinic acid, (DMSA, Aldrich). These particles averaged 10–12 nm in diameter as determined by transmission electron microscopy. Other properties of these NPs are reported in Kim et al.[56]

NPs were added to hydrogels from the aqueous dispersions of FePt and $CoFe_2O_4$ developed here.

7.2.5 Hydrogel Synthesis

Copolymer hydrogels of N-isopropylacrylamide (NIPAAm, Acros Organics, Fair Lawn, NJ) and acrylamide (AAm, Acros) were formed by free radical solution polymerization using a 1:1 mixture of methanol (Acros):water as the solvent. MNPs were added at this step, with aqueous MNP solutions (FePt, $CoFe_2O_4$, or γ-Fe_2O_3) replacing the water. The NIPAAm:AAm ratios were adjusted to develop materials with LCSTs above physiological temperature. After mixing the monomers with the solvent, the crosslinking agent methylene bisacrylamide (MBAAm, Aldrich) was added at between 1 and 10 mol % of the monomer content. This solution was mixed thoroughly while being bubbled with nitrogen to remove dissolved oxygen. Finally, redox initiator ammonium persulfate (AMPS, Acros) and reaction accelerator N,N,N',N'-tetramethylethylenediamine (TEMED, ICN Biomedics, Inc.) were added to the mixture and dissolved. The mixture was placed between siliconized glass plates separated by a Teflon® spacer to control sample thickness. The reaction was carried out at room temperature for 24 hours.

A nonthermally responsive hydrogel was used for magnetic heating experiments. Here, polymer gels of 2-hydroxyethyl methacrylate (Acros), or PHEMA, were synthesized using deionized (DI) water as the solvent, MBAAm as the crosslinking agent, and AMPS combined with sodium metabisulfite (Acros), both added at 1 wt % of the monomer, as the redox initiator system.

After formation, hydrogels were recovered from the glass plates and disc-shaped samples were cut out using a cork borer, yielding samples with approximately 10 mm diameter and less than 1 mm thickness. The samples were rinsed thoroughly in water over several days (replacing the water frequently) to extract any unreacted monomer.

7.2.6 MNP Retention in Hydrogels

When freshly formed hydrogels were placed in water to extract unreacted monomer, the loss of any MNPs from the structures was determined by UV/Vis spectrophotometry (Shimadzu, Model UV-2401PC, Norcross, GA) by measuring an aliquot of the supernatant at 500 nm and comparing to a calibration curve created made using known concentrations of MNPs. MNP-loaded hydrogels were also examined using optical microscopy to determine whether the NPs agglomerated during the polymerization.

7.2.7 Equilibrium Swelling

Swelling experiments were used to determine the thermosensitivity of the P(NIPAAm-co-AAm) hydrogels, with and without MNPs. Equilibrium swelling experiments were conducted by first drying the disc-shaped hydrogel samples to constant weight under vacuum to determine the dry weight. Then the samples were placed in DI water at fixed temperatures from 25°C to 60°C and allowed to equilibrate for at least 3 days at each temperature. The wet weights were found by removing each gel sample from the water, and blotting the surface with a Kimwipe® and weighing. The equilibrium polymer fractions, which are the inverse of the swelling ratios (1/q), were calculated at each temperature as the ratio of the dry weight to the wet weight of each sample.

7.2.8 Drug Loading

Two drugs were investigated as representative of smaller molecular weight biologically active agents: theophylline (Sigma), and 5-fluorouracil (5-FU, Sigma). The drugs were loaded by equilibrium partitioning into hydrogel discs after they had been thoroughly rinsed. Gels were soaked in concentrated aqueous solutions of theophylline or 5-FU for at least 5 days and equilibrated at the same temperature used at the beginning of the drug release experiment.

7.2.9 Constant Temperature and Pulsatile Drug Release

Drug release experiments were conducted either using a constant-temperature water bath or by changing the temperature of the release medium to mimic magnetothermal activation. For the constant temperature drug release experiments, a six-chambered Type II USP dissolution system (Distek, Model 2100C, North Brunswick, NJ) was used. The dissolution cells were filled with DI water and the bath temperature equilibrated prior to beginning the experiment. At time zero of the experiment, one drug-loaded sample disc (that had been rinsed briefly to remove drug bound to the surface during the equilibrium partitioning loading) was placed in each cell. The absorbance of the release media were monitored as a function of time using a set of flow-through cuvettes with a peristaltic pump to continuously sample each release cell (this flow was returned to the dissolution cell, thus maintaining a constant release volume during the experiment). The concentration (and mass) of drug released was determined using a calibration curve generated according to Beer's law for known concentrations of both theophylline and 5-FU where the absorbances were measured at 269 nm for both model drugs.

Pulsatile release experiments were conducted using 200 mL of DI water placed in glass jars inside shaking water baths at 37°C and 47°C. At time zero of each experiment, a disc-shaped sample freshly loaded by equilibrium partitioning were briefly rinsed and blotted with a Kimwipe® and placed in the first 200 mL release medium. At every 30 minutes of the release period, each sample was transferred to a jar at the other temperature, resulting in a temperature profile of the medium that shifted between 37°C and 47°C every 30 minutes. Release rates were monitored by removing a 3-mL aliquot of sample every 5 minutes and measuring absorbance at 269 nm. Calculations for mass of drug released took into consideration the decrease in release volume due to the removal of the sample aliquots.

7.2.10 Magnetic Heating

PHEMA gels loaded with 0.75 wt % FePt MNPs were used for magnetic heating studies. The gels were preequilibrated in water and exposed to magnetic fields in a six-turn solenoid coil using a custom-built magnetic-induction hyperthermia chamber (Induction Atmospheres, Rochester, NY), which generates magnetic fields from 100–700 Gauss with frequencies in the range of 50–450 kHz. The instrument has a 5 kW power supply (Ameritherm, Inc., Scottsville, NY) and a connected circulating chiller bath (Koolant Koolers®, Inc., Kalamazoo, MI). More details on the magnetic heating system can be found in a paper by Kim et al.[56] The gels were subjected to a magnetic field for 15 minutes. Temperatures were monitored using an infrared thermal imaging camera (FLIR Thermacam® SC2000, FLIR Systems, Boston, MA).

7.3 Results and Discussion

7.3.1 Magnetic Capture of Aqueous-Dispersed NPs

Effective magnetic capture relies on NP dispersions being subjected to a static magnetic field of sufficient strength over a period to overcome the particle momentum as MNPs pass through a simulated blood vessel. Using linear flow rates similar to that which would be expected in normal vascular flow, the capture of particles was determined as a function of magnetic field strength (Figure 7.2). Raw data of absorbance vs. time were recorded using the spectrophotometer, with the reduction in absorbance used to determine the percentage of NPs that were captured in each experiment. In Figure 7.2, maghemite NPs were captured at over 60% for the slower flow rate of 1 cm/s, for each of magnetic field strengths tested. In this case, many of the NPs did not have sufficient momentum to escape the magnetic field. When a higher linear velocity was used (3 cm/s), the difference in capture due to

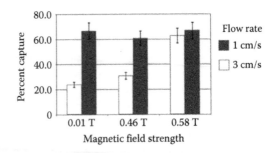

FIGURE 7.2
Effect of flow velocity and magnetic field strength on capture of maghemite nanopowder dispersed in water using sulfosuccinic acid. Tube diameter = 1.0 mm, MNP concentration = 0.05 g/L. Error bars represent the standard deviation for three replicates.

applied field strength is more pronounced. The highest field strength (0.58 T) could still capture over 60% of the particles, but lower field strengths led to a reduced percent capture. Of course, for applications *in vivo*, the capture efficiency is also dependent on the depth of blood vessels under the skin, since the field strength decreases sharply with distance from a static magnet.

7.3.2 Effect of Tubing Diameter and Viscosity Modifiers on Magnetic Capture

With larger tube diameters (while maintaining a fixed linear velocity), magnetic capture was reduced (Figure 7.3). For aqueous solutions of maghemite, increasing the vessel diameter from 1–6 mm had a significant impact on magnetic capture, even though the magnetic field was applied so that the measured field strength would reach the centerline of the vessel. This reduction (which is also noted for solutions containing PEG as a viscosity modifier) is explained by two factors: (1) the increased mass flow rate in the larger diameter tubes, even though the linear flow rate was kept constant between experiments and (2) the magnetic field, which was applied from one side of the tubing dropped considerably before reaching across the entire cross section of the larger tubes.

While the capture of MNPs was somewhat reduced by the addition of PEG (which acted to sweep the NPs through the capture region), the addition of blood proteins albumin and γ-globulin caused a marked decrease in

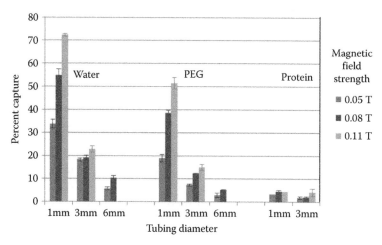

FIGURE 7.3

Effect of solution viscosity on capture of maghemite nanopowder dispersed in water using sulfosuccinic acid. MNP concentration = 0.05 g/L, Flow rate = 3.3 cm/s, poly(ethylene glycol) (PEG) molecular weight = 8000, PEG concentration = 4.8 wt %, PEG solution kinematic viscosity = 3.64 mm^2/s. Protein solution comprised of 46.5 g/L albumin and 63.5 g/L γ-globulins, protein solution kinematic viscosity = 1.74 mm^2/s. There are no data for capture in 6-mm tube diameters for 0.11 T magnetic fields, as fields of this strength could not be generated at the centerline of the vessels using the neodymium magnets available.

capture efficiency, to less than 5%, regardless of the field strength applied (up to 0.11 T; Figure 7.3). Two key factors in this reduced capture efficiency are protein molecular weight and opsonization. For the first factor, despite the PEG solution and protein solution having similar kinematic viscosities, the proteins used had much higher molecular weights than the PEG molecules (66,000 and 120,000 for albumin and γ-globulin, respectively, compared to 8000 for PEG). This higher molecular weight caused greater flow momentum, preventing particles from being held in place by the magnetic field. The second factor which is commonly referenced for foreign materials placed in the bloodstream is opsonization.[13] Blood proteins are part of the immune response system, and they act to remove unnatural materials from the blood stream by binding to the surface and beginning the process of phagocytosis. In this experiment, only the proteins were present, but they naturally attach to the surface of MNPs when in aqueous solution, rendering the relatively small NPs (~30–50 nm) much larger and causing aggregations. The NP aggregates possess considerable momentum when flowing in the capture zone, reducing capture efficiency several fold compared to the experiments in aqueous solution. To prevent opsonization, coatings with PEG are known to reduce the rate of protein adhesion,[57] which would allow individual particles to be captured by static magnets.

7.3.3 Effect of Protein Opsonization

Extending magnetic capture from protein solutions to blood plasma, FBS was used to disperse the maghemite NPs, with capture experiments conducted using 0.01 T magnetic fields (Figure 7.4). The results here mirror those seen in Figure 7.3, where (in this case) the proteins present in FBS caused a marked decrease in percent capture. Magnetic capture trends also paralleled those in Figure 7.4 with respect to vessel diameter, although for the 0.5 mm diameter tubing, there were some protein/NP aggregates which affected flow after periods of time, likely due to adhesion to the wall of the tube followed by blockage of flow.

To investigate the opsonization process, MNP dispersions were added to a dilute solution of albumin and globulins, with DLS used to determine the hydrodynamic diameter at different times after addition. As a control, the protein solution showed only a minor increase in average size over a 2-hour window (Figure 7.5, top curves). By comparison, when the proteins were mixed with the maghemite NPs in water, large aggregates appeared after only 5 minutes (Figure 7.5, middle curves). Since PEG and other polymer coatings can improve dispersions while also minimizing opsonization, a similar experiment was conducted using commercial fluidMAG PAD, which consists of magnetite coated with polyacrylamide (Figure 7.5, bottom curves). In this case, there was no significant change in the particle size distribution, even over 72 hours after exposure to the albumin and globulin solution.

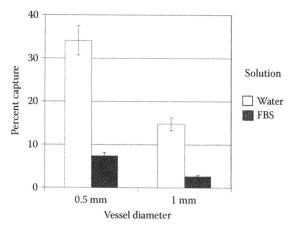

FIGURE 7.4

Comparison of magnetic capture of maghemite nanopowder dispersed using sulfosuccinic acid in water or fetal bovine serum (FBS). MNP concentration = 0.05 g/L, magnetic field strength at vessel centerline = 0.01 T, flow velocity = 3 cm/s. Error bars represent the standard deviation for three replicates.

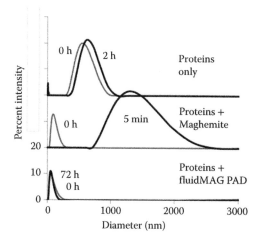

FIGURE 7.5

Particle size distributions of dispersed maghemite nanopowder solutions before and after exposure to protein solution (0.30 g/L albumin and 0.15 g/L α- and γ-globulins). Size distributions were determined using dynamic light scattering, with number average size distributions reported.

The results presented above show that achieving successful magnetic capture in biological environments requires optimization. While capture in aqueous solutions is certainly feasible, stopping NPs in flow that includes proteins is significantly reduced. Without a polymer coating to prevent or slow the opsonization process, protein adhesion to NPs causes a large increase in the hydrodynamic diameter of each flowing MNP, leading to greater particle momentum that cannot be held in place using static magnets (at least using the fairly strong neodymium magnets investigated here; and these magnets present their own hazards for handling due to their strong attraction).

7.3.4 Magnetic Hydrogels: MNP Synthesis and Dispersion

In addition to the maghemite NPs used for magnetic capture, two other MNPs were synthesized for study in hydrogels. FePt and $CoFe_2O_4$ MNPs were successfully synthesized and dispersed in aqueous solution after a ligand-exchange reaction to replace oleic acid used in the organic synthesis with either merceptoundecanoic acid (for FePt) or DMSA (for $CoFe_2O_4$) surfactants.[56] Using transmission electron microscopy, the FePt MNPs were found to have uniform diameters of 4–5 nm[54] and the $CoFe_2O_4$ NPs of approximately 7 nm.[56] The $CoFe_2O_4$ MNPs were found to form aggregates with an intensity-average hydrodynamic diameter of 234 nm determined by DLS.[54]

7.3.5 Gel Synthesis

P(NIPAAm-co-AAm) and PHEMA hydrogels formed uniform, clear gels. With the addition of MNPs to the gels, a gradient in color was observed, with the MNPs imparting a light brown appearance to the gels (Figure 7.6). In some of the gels with high concentrations of MNPs (or if an unstable aqueous dispersion were used), significant aggregation of MNPs in PHEMA gels was observable under an optical microscope, with some aggregates as large as 100 μm. By thorough sonication and adjustment of surfactant quantities and control of pH, hydrogels were synthesized with well-dispersed MNPs and no agglomerations observable under an optical microscope.

FIGURE 7.6
P(NIPAAm-co-16.6%AAm) gels containing various amounts of FePt. From left to right, the samples contain: 0 wt %, 0.002 wt %, 0.05 wt %, and 0.125 wt % FePt. Gel disks are approximately 1 cm in diameter.

7.3.6 MNP Retention in Gels

For MNP-loaded hydrogels to be used for magnetothermally triggered drug delivery, it is vital that the MNPs are trapped inside the hydrogels, so a mass balance was conducted on the wash solution to determine the fraction of MNPs that were rinsed from the hydrogels. Over 99% of the MNPs were successfully retained in the PHEMA gels, because, even though the MNPs were quite small (5–10 nm, or 50–100 Å), they were slightly larger than the average mesh space available for diffusion in the hydrogels (typically around 50 Å for PHEMA[45]). The small amount of MNPs that were observed in the wash solution are believed to have desorbed from the hydrogel surfaces upon immersion in water, especially since the absorbance of the wash solution rose quickly, but no increase was observed over the next several days of washing.

7.3.7 Equilibrium Swelling

By investigating the temperature-dependent equilibrium swelling behavior of the P(NIPAAm-co-AAm) hydrogels, we could select candidate materials with LCST values slightly higher than physiological temperature (37°C), which were further investigated for hyperthermia-activated squeezing drug release. To achieve a hydrogel with an LCST over 40°C, a 83.4:16.6 molar ratio of NIPAAm to AAm was used, as described elsewhere[58]); however, 90:10 NIPAAm:AAm gels were also investigated to achieve a sharper transition at temperature above 40°C, even for gels with higher crosslink ratios. As a side note, while it is customary to describe the composition of hydrogels based on the feed ratio of monomers with single polymerizable double bonds, crosslink ratios over a few percent have a marked effect on properties, as a difunctional monomer, such as methylene bisacrylamide changes the character of the network. The LCST and degree of swelling are dependent on polymer composition as well as the crosslink ratio (see Figure 7.7, where the swelling results are shown as polymer fractions or $1/q$). The hydrogel with 83.4% NIPAAm and a 1 mol % crosslink ratio, displayed a sharp change in swelling between 37°C and 47°C; however, in the gels with additional MBAAm crosslinker, a much less-pronounced temperature-induced phase separation was noted. This is due to a combination of the additional crosslinking agent restricting polymer expansion as well as the high concentrations of MBAAm acting as a comonomer which interferes with the cooperative thermal transition in LCST polymers. In the hydrogels with 90% NIPAAm, a more gradual transition was observed between 37°C and 47°C, but this transition remained prevalent, even with the higher crosslink ratio (5 mol % MBAAm). Thus, in drug delivery systems where a larger solute is used, the lightly (1 mol %) crosslinked P(NIPAAm-co-16.6% AAm) gel would be preferred, while for small solutes (such as 5-FU and theophylline used in this work), the 5 mol % P(NIPAAm-co-10% AAm) was expected to perform better.

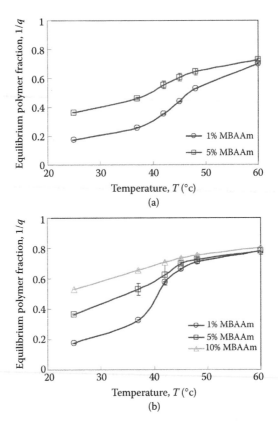

FIGURE 7.7
Equilibrium swelling behavior of (a) P(NIPAAm-co-16.6%AAm) and (b) P(NIPAAm-co-10%AAm) as a function of temperature for different crosslinking ratios. The error bars represent the standard deviation for three replicates. In some cases, the error bars are smaller than the data symbols.

When MNPs are incorporated into hydrogels, they can influence swelling behavior. Because the MNPs were added during the free radical solution polymerization step, the MNPs could interfere with the formation of chains and crosslinks during the reaction, resulting in fewer successful crosslinks (and higher swelling ratios). However, this is counterbalanced by the additional unswellable dry weight that the MNPs add to the hydrogels, and the result is that the swelling of the P(NIPAAm-co-AAm) gels with 83.4% NIPAAm was largely unaffected by the presence of FePt MNPs or a model drug, theophylline. The temperature dependence of weight swelling ratios for gel containing 0.05 wt % FePt and 1.0 wt % theophylline were not statistically different from gels containing no MNPs or drug, with equilibrium swelling ratios being 4.5 at 37°C and 1.5 at 50°C. Thus, the MNPs and theophylline were not used at concentrations high enough to have a significant

impact on hydrogel structure and importantly, the MNPs did not interfere with the hydrophilic-to-hydrophobic transition characteristic of LCST
polymers.

7.3.8 Drug Release

For drug-release experiments, the two primary goals were to examine the
impact of the addition of FePt MNPs on release kinetics and to study the
thermally controlled pulsatile release of two bioactive agents: theophylline
and 5-fluorouracil.

Temperature-sensitive drug release was investigated by running constant temperature experiments at 37°C and 50°C for theophylline release
from initially dry P(83.4%NIPAAm-co-16.6%AAm) hydrogels with various
concentrations of FePt MNPs. The release curves were essentially undistin-
guishable, with the addition of MNPs to the gels causing a slight increase
in the release rate, at least up to 0.125 wt % FePt, the highest concentration
investigated. While constant temperature release does not demonstrate thermally activated release, it offers an important confirmation that the presence
of a small amount of MNPs in the gels (which are necessary for magnetothermal activation) did not have an appreciable effect on the gel structure. A
further set of constant temperature release experiments were conducted at
50°C, and the diffusion coefficients of theophylline were estimated from the
early-time release data, using[59]:

$$\left(\frac{M_t}{M_\infty}\right) = 4\left(\frac{Dt}{\pi\delta^2}\right)^{\frac{1}{2}} \text{ for } 0 \le \left(\frac{M_t}{M_\infty}\right) \le 0.6 \tag{7.1}$$

where M_t is the amount of drug released at time t, M_∞ is the amount
released at infinite time, δ is the half-thickness of the sample, t is time, and
D is the diffusion coefficient (Table 7.2). Here, the diffusion coefficients

TABLE 7.2

Diffusion Coefficients for Theophylline Release from P(NIPAAm-co-16.6%
AAm) Hydrogels

FePt MNP Content (wt %)	Diffusion Coefficient ($\times 10^8$ cm²/s)	
	T = 37°C	**T = 50°C**
0	4.79 ± 0.09	3.28 ± 0.28
0.002	4.47 ± 0.20	3.37 ± 0.15
0.050	4.35 ± 0.08	4.09 ± 0.36
0.125	4.06 ± 0.21	3.10 ± 0.24

Note: Experimental error is reported as the standard deviation for three replicates.

were lower at 50°C (except for the gels with 0.05 wt % FePt, where no statistical difference was noted), indicating that the gels did not swell to as great an extent at the higher temperature. The difference in diffusion coefficients was appreciable, but they were only about 25% lower at the raised temperature, most likely due to the small molecular size and hydrophilicity of theophylline, which was able to readily diffuse through the polymer mesh.

To mimic the magnetothermal triggering of release, where gels originally at physiological temperature are heated by an AC magnetic field, pulsatile release experiments were done using P(NIPAAm-co-AAm) hydrogels (containing either 83.4% or 90 mol % NIPAAm monomer in the reaction feed) by starting release using gels equilibrated in an aqueous theophylline loading solution. After briefly rinsing the gels to remove surface-bound drug, the samples were placed in aqueous release media alternating between 37°C and 48°C for 30-minute intervals.

Release of theophylline from P(NIPAAm-co-AAm) gels with 83.4 mol % NIPAAm and 1 mol % crosslinking showed a response to temperature changes (Figure 7.8a), with an increase in theophylline release upon heating (squeezing release), but also upon cooling (just after 60 minutes) as the gel expanded at the cooler temperature. For the more highly crosslinked gel with 83.4% NIPAAm, the release profile was unaffected by the temperature change of the release medium. This is due largely to the less-sharp LCST observed for the gels containing the higher proportion of MBAAm, which allowed a much smaller swelling response between 37°C and 48°C compared to the gel with 1 mol % MBAAm (as seen in Figure 7.7a). Theophylline release from poly(NIPAAm-co-10%AAm) gels showed the similar results (Figure 7.8b). The 1 mol %-crosslinked gel with 10% AAm had a less-pronounced pulsatile release, but release from the gels with 5 mol % MBAAm crosslinker had a similar profile to Figure 7.8a, indicating that even though the LCST was tuned appropriately, the release of theophylline could not be controlled by changes in the polymer structure due to the pulsatile temperature of the release medium.

A second model drug, 5-fluorouracil (5-FU), which is used in cancer therapy, was investigated for temperature-triggered release (Figure 7.9). While drug release from the gels with 83.4 mol % NIPAAm showed little sensitivity to the temperature changes, the 5-FU release profiles from gels with 90 mol % NIPAAm showed squeezing release behavior, as the 5-FU was released in short bursts upon heating above the LCST of these gels. For both of these hydrogel compositions, the crosslink ratio had little effect on the overall release profile. Additionally, aside from the initial period, 5-FU release in the second 37°C period (from 60 to 90 minutes) was nearly flat, indicating that this system could be used for magnetothermally triggered release of 5-FU, where the activation of the drug release is caused by magnetic heating.

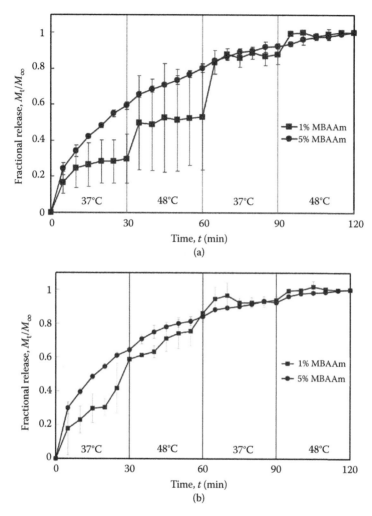

FIGURE 7.8
Thermally pulsed theophylline release from (a) P(NIPAAm-co-16.6%AAm) and (b) P(NIPAAm-co-10%AAm) hydrogels. The gels were equilibrated in theophylline loading solution at 37°C prior to the release experiment. Error bars represent the standard deviation for three replicates.

7.3.9 Magnetic Heating

Because the design of a magnetothermal-responsive system requires that the heating come from a magnetic source rather than a temperature-controlled water bath, a magnetic heating experiment was conducted on a disc-shaped equilibrium-swollen PHEMA gel with imbedded FePt MNPs. The gel was successfully heated using an AC field to reach a maximum temperature of

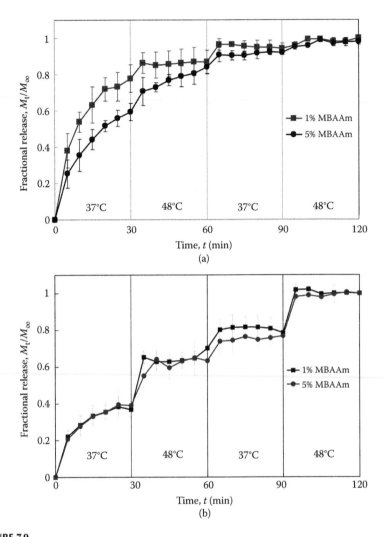

FIGURE 7.9
Thermally pulsed release of 5-fluorouracil from (a) P(NIPAAm-co-16.6%AAm) and (b) P (NIPAAm-co-10%AAm) at 37°C and 48°C. The gels were equilibrated in 5-FU loading solution at 37°C prior to the release experiment. Error bars represent the standard deviation for three replicates.

31°C on the gel surface in less than 10 minutes, starting from room temperature (Figure 7.10). The heating was observed to be uniform, as readings were taken from three radial positions (at r = 0, R/2, and R, where R is the radius of the disc). Fine-tuning of the MNP and magnetic field properties will enable the system to heat from physiological conditions to hyperthermic temperatures (above 42°C), as has been shown elsewhere using iron oxide, manganese ferrite, and cobalt ferrite MNPs.[44,56,60]

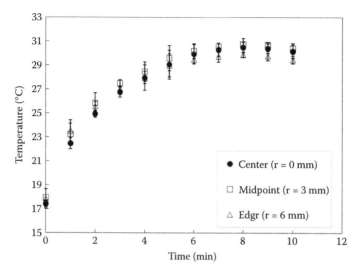

FIGURE 7.10
Temperature profiles for a PHEMA gel containing 0.75 wt % FePt MNPs subjected to a 560 Oe AC magnetic field at 231 kHz. Temperature readings were taken at $r = 0$, R/2, and R on a disc-shaped sample with radius R using an infrared camera. Error bars represent the standard deviation for three heating experiments.

7.4 Conclusions

Magnetic capture and magnetothermal-triggered delivery are two therapeutic enhancement methods that utilize magnetic actuation. Both are the focus of numerous papers and show potential for developing more sophisticated, better-targeted medicine. However, significant challenges must be overcome to realize the potential benefits to patients.

Successful magnetic capture relies on the development of MNPs, which can be stably dispersed in blood or other aqueous environments and avoid opsonization. When the momentum of the particles is larger, such as when the particles are encased in proteins, capture efficiency can be too low to be effective. Stronger static magnets can be most effective at collecting and localizing NPs in capillaries with slower flow rates that are near the skin, as the magnetic field intensity drops precipitously with distance. While solution viscosity has an effect on capture, protein opsonization is perhaps the greatest challenge that must be overcome in collecting MNPs for targeted therapy using magnetic capture.

The development of magnetothermally triggered delivery systems requires the design of both magnetic nanomaterials that can be heated by application of an AC magnetic field and thermally responsive polymers that can sequester a drug at physiological temperatures but release rapidly upon magnetic heating. Here, we have successfully synthesized thermally responsive materials that exhibit phase separation above 37°C, and showed that pulsatile release of 5-FU could be achieved upon heating the polymer to 48°C. We have also seen that hydrogels loaded with as little as 0.75 wt % FePt MNPs could be heated by an AC magnetic field, although more concentrated (or different) MNPs, or the tuning of the applied magnetic field intensity and strength are needed to increase the heating to reach hyperthermic temperatures. An additional challenge to working with MNPs loaded at high concentrations is their tendency to aggregate into micron-sized agglomerates; this was not observed to affect the ability of MNPs to heat (or at even larger concentrations reported by Satarkar and Hilt),[33] but may impact the uniformity of heating within a hydrogel. Importantly, the MNPs were observed to remain trapped inside the hydrogel networks, and not have a significant influence on drug diffusion and release behavior. Further work will focus on developing polymeric materials with a sharp LCST transition so that heating to hyperthermia conditions (42°C) can trigger drug release. Drug release prior to triggering (e.g., the first 30 minutes of the pulsatile release profiles shown) should also be minimized so that potent chemotherapy drugs can be used effectively while reducing the potential for the side effects of the drug. Continued work will also focus on combining the thermoresponsive materials with MNPs so that release is actuated by a magnetic field.

Magnetic actuation using static or alternating magnetic fields holds promise for advanced therapeutic approaches, but magnetic capture and magnetothermally triggered drug release both require improvements to optimize material properties and functionality.

Acknowledgments

The authors acknowledge support through the University of Alabama's Alton Scott Memorial Award, McWane fellowships to M. L. Hampel, M. K. Sewell, and A. E. Frees, a Sigma Xi Undergraduate Research Grant-in-Aid to A. E. Frees, an NSF REU Grant (#1062611) to support L. M. Blue, N. Lapp, M. Zhang, and J. M. Robertson, an R21 grant from the National Cancer Institute (NIH Grant R21CA 141388), a Fulbright Distinguished Scholarship to C. S. Brazel, and University of Alabama's Department of Chemical and Biological Engineering. The authors also appreciate the assistance and valuable discussions with D. E. Nikles, D.-H. Kim, I. Ankareddi, and H. G. Bagaria at the University of Alabama.

References

1. L. Brannon-Peppas, J. O. Blanchette, Nanoparticle and targeted systems for cancer therapy, *Adv. Drug Deliv. Rev.*, **56** (2004) 1649–1659.
2. I. Brigger, C. Dubernet, P. Couvreur, Nanoparticles in cancer therapy and diagnosis, *Adv. Drug Deliv. Rev.*, **54** (2002) 631–651.
3. D.-H. Kim, H. Zeng, T. C. Ng, C.S. Brazel, T1 and T2 relaxivities of succimer-coated $MFe_2^+O_4$ (M = Mn^{2+}, Fe^{2+} and Co^{2+}) inverse spinel ferrites for potential use as phase-contrast agents in medical MRI, *J. Magn. Magn. Mater.*, **321** (2009) 3899–3904.
4. Q. A. Pankhurst, J. Connolly, S. K. Jones, J. Dobson, Applications of magnetic nanoparticles in biomedicine, *J. Phys. D Appl. Phys.*, **36** (2003) R167.
5. C. S. Brazel, Magnetothermally-responsive nanomaterials: Combining magnetic nanostructures and thermally-sensitive polymers for triggered drug release, *Pharm. Res.*, **26** (2009) 644–656.
6. N. J. Darton, B. Hallmark, X. Han, S. Palit, N. K. H. Slater, M. R. Mackley, The in-flow capture of superparamagnetic nanoparticles for targeting therapeutics, *Nanomedicine*, **4** (2008) 19–29.
7. B. Thiesen, A. Jordan, Clinical applications of magnetic nanoparticles for hyperthermia, *Int. J. Hyperthermia*, **24** (2008) 1–8.
8. T. M. Fahmy, P. M. Fong, A. Goyal, W. M. Saltzman, Targeted for drug delivery, *Mater. Today*, **8** (2005) 18–26.
9. Y. Lu, P. S. Low, Folate-mediated delivery of macromolecular anticancer therapeutic agents, *Adv. Drug Deliv. Rev.*, **54** (2002) 675–693.
10. J. S. Desgrosellier, D. A. Cheresh, Integrins in cancer: Biological implications and therapeutic opportunities, *Nat. Rev. Cancer*, **10** (2010) 9–22.
11. R. van der Meel, L. J. C. Vehmeijer, R. J. Kok, G. Storm, E. V. B. van Gaal, Ligand-targeted particulate nanomedicines undergoing clinical evaluation: Current status, *Adv. Drug Deliv. Rev.*, **65** (2013) 1284–1298.
12. K. H. Min, H. J. Lee, K. Kim, I. C. Kwon, S. Y. Jeong, S. C. Lee, The tumor accumulation and therapeutic efficacy of doxorubicin carried in calcium phosphate-reinforced polymer nanoparticles, *Biomaterials*, **33** (2012) 5788–5797.
13. D. E. Owens III, N. A. Peppas, Opsonization, biodistribution, and pharmacokinetics of polymeric nanoparticles, *Int. J. Pharm.*, **307** (2006) 93–102.
14. W. Tan, H. Wang, Y. Chen, X. Zhang, H. Zhu, C. Yang, R. Yang, C. Liu, Molecular aptamers for drug delivery, *Trends Biotechnol.*, **29** (2011) 634–640.
15. V. A. Petrenko, P. K. Jayanna, Phage protein-targeted cancer nanomedicines, *FEBS Lett.*, **588** (2014) 341–349.
16. P. Urban, J. Estelrich, A. Cortes, X. Fernandez-Busquets, A nanovector with complete discrimination for targeted delivery to plasmodium falciparum-infected versus non-infected red blood cells in vitro, *J. Control. Release*, **151** (2011) 202–211.
17. B. Hallmark, N. J. Darton, T. James, P. Agrawal, N. K. H. Slater, Magnetic field strength requirements to capture superparamagnetic nanoparticles within capillary flow, *J. Nano. Res.*, **12** (2010) 2951–2965.
18. A. E. David, A. J. Cole, B. Chertok, Y. S. Park, V. C. Yang, A combined theoretical and in vitro modeling approach for predicting the magnetic capture and retention of magnetic nanoparticles in vivo, *J. Control. Release*, **152** (2011) 67–75.

19. B. Chertok, B. A. Moffat, A. E. David, F. Q. Yu, C. Bergemann, B. D. Ross, V. C. Yang, Iron oxide nanoparticles as a drug delivery vehicle for MRI monitored magnetic targeting of brain tumors, *Biomaterials*, **29** (2008) 487–496.

20. N. Doshi, Flow and adhesion of drug carriers in blood vessels depend on their shape: A study using model synthetic microvascular networks. *J. Control. Release*, **146** (2010) 196–200.

21. P. Decuzzi, Size and shape effects in the biodistribution of intravascularly injected particles, *J Control. Release*, **141** (3) (2010) 320–327.

22. A. J. Rosengart, M. D. Kaminski, H. Chen, P. L. Caviness, A. D. Ebner, J. A. Ritter, Magnetizable implants and functionalized magnetic carriers: A novel approach for noninvasive yet targeted drug delivery, *J. Magn. Magn. Mater.*, **293** (2005) 633–638.

23. R. Ivkov, S. J. DeNardo, W. Daum, A. R. Foreman, R. C. Goldstein, V. S. Nemkov, G. L. DeNardo, Application of high amplitude alternating magnetic fields for heat induction of nanoparticles localized in cancer, *Clin. Cancer Res.*, **11** (2005) 7093s–7103s, doi:10.1158/1078-0432.CCR-1004-0016.

24. M. Gonzales-Weimuller, M. Zeisberger, K. M. Krishnan, Size-dependant heating rates of iron oxide nanoparticles for magnetic fluid hyperthermia, *J. Magn. Magn. Mater.*, **321** (2009) 1947–1950.

25. X. Yin, A. S. Hoffman, P. S. Stayton, Poly(N-isopropylacrylamide-co-propylacrylic acid) copolymers that respond sharply to temperature and pH, *Biomacromolecules*, **7** (2006) 1381–1385.

26. K. H. Bae, S. H. Choi, S. Y. Park, Y. Lee, T. G. Park, Thermosensitive pluronic micelles stabilized by shell cross-linking with gold nanoparticles, *Langmuir*, **22** (2006) 6380–6384.

27. S. Sershen, J. West, Implantable, polymeric systems for modulated drug delivery, *Adv. Drug Deliv. Rev.*, **54** (2002) 1225–1235.

28. M. Bikram, A. M. Gobin, R. E. Whitmire, J. L. West, Temperature-sensitive hydrogels with SiO_2-Au nanoshells for controlled drug delivery, *J. Control. Release*, **123** (2007) 219–227.

29. G. A. Husseini, N. Y. Rapoport, D. A. Christensen, J. D. Pruitt, W. G. Pitt, Kinetics of ultrasonic release of doxorubicin from pluronic P105 micelles, *Colloids Surf. B Biointerfaces*, **24** (2002) 253–264.

30. M. Saravanan, K. Bhaskar, G. Maharajan, K. S. Pillai, Ultrasonically controlled release and targeted delivery of diclofenac sodium via gelatin magnetic microspheres, *Int. J. Pharm.*, **283** (2004) 71–82.

31. J. T. Santini, M. J. Cima, R. Langer, A controlled-release microchip, *Nature*, **397** (1999) 335–338.

32. A. Nisar, N. Afzulpurkar, B. Mahaisavariya, A. Tuantranont, MEMS-based micropumps in drug delivery and biomedical applications, *Sensors Actuat. B Chem.*, **130** (2008) 917–942.

33. N. S. Satarkar, J. Z. Hilt, Magnetic hydrogel nanocomposites for remote controlled pulsatile drug release, *J. Control. Release*, **130** (2008) 246–251.

34. V. Saini, V. P. Zharov, C. S. Brazel, D. E. Nikles, D. T. Johnson, M. Everts, Combination of viral biology and nanotechnology: New applications in nanomedicine, *Nanomedicine*, **2** (2006) 200–206.

35. R. Hergt, S. Dutz, R. Muller, M. Zeisberger, Magnetic particle hyperthermia: Nanoparticle magnetism and materials development for cancer therapy, *J. Phys. Condens. Matter*, **18** (2006) S2919–S2934.

36. P. Wust, U. Gneveckow, M. Johannsen, D. Bahmer, T. Henkel, F. Kahmann, J. Sehouli, R. Felix, J. Ricke, A. Jordan, Magnetic nanoparticles for interstitial thermotherapy—Feasibility, tolerance and achieved temperatures, *Int. J. Hyperthermia*, **22** (2006) 673–685.
37. A. Ito, Y. Kuga, H. Honda, H. Kikkawa, A. Horiuchi, Y. Watanabe, T. Kobayashi, Magnetite nanoparticle-loaded anti-HER2 immunoliposomes for combination of antibody therapy with hyperthermia, *Cancer Lett.*, **212** (2004) 167–175.
38. S. B. Lee, S. H. Lee, D. H. Kim, D. Y. Lee, Y. K. Lee, K. N. Kim, K. M. Kim, In vitro cytotoxicity of alginate-encapsulating ferrite particles using WST-1, *Bioceramics*, **17** (2005) 815–818.
39. A. Jordan, R. Scholz, P. Wust, H. Fahling, R. Felix, Magnetic fluid hyperthermia (MFH): Cancer treatment with AC magnetic field induced excitation of biocompatible superparamagnetic nanoparticles, *J. Magn. Magn. Mater.*, **201** (1999) 413–419.
40. D. P. O'Neal, L. R. Hirsch, N. J. Halas, J. D. Payne, J. L. West, Photo-thermal tumor ablation in mice using near infrared-absorbing nanoparticles, *Cancer Lett.*, **209** (2004) 171–176.
41. R. Hergt, W. Andra, C.G. d'Ambly, I. Hilger, W. A. Kaiser, U. Richter, H. G. Schmidt, Physical limits of hyperthermia using magnetite fine particles, *IEEE Trans. Magn.*, **34** (1998) 3745–3754.
42. M. Johannsen, U. Gneveckow, K. Taymoorian, B. Thiesen, N. Waldoefner, R. Scholz, K. Jung, A. Jordan, P. Wust, S. A. Loening, Morbidity and quality of life during thermotherapy using magnetic nanoparticles in locally recurrent prostate cancer: Results of a prospective phase I trial, *Int. J. Hyperthermia*, **23** (2007) 315–323.
43. R. E. Rosensweig, Heating magnetic fluid with alternating magnetic field, *J. Magn. Magn. Mater.*, **252** (2002) 370–374.
44. D.-H. Kim, Y. Thai, D. E. Nikles, C. S. Brazel, Optimized heat generation of $MnFe_2O_4$ nanoparticles for magnetic hyperthermia using multifunctional nanoparticles, *IEEE Trans. Magn.*, **45** (2009) 64–70.
45. I. Ankareddi, C. S. Brazel, Synthesis and characterization of grafted thermosensitive hydrogels for heating activated controlled release, *Int. J. Pharm.*, **336** (2007) 241–247.
46. C. S. Brazel, N. A. Peppas, Pulsatile local delivery of thrombolytic and anti-thrombotic agents using poly(N-isopropylacrylamide-co-methacrylic acid) hydrogels, *J. Control. Release*, **39** (1996) 57–64.
47. A. Hoffman, Applications of thermally reversible polymers and hydrogels in therapeutics and diagnostics, *J. Control. Release*, **6** (1987) 297–305.
48. A. Kikuchi, T. Okano, Pulsatile drug release control using hydrogels, *Adv. Drug Deliv. Rev.*, **54** (2002) 53–77.
49. D. Schmaljohann, Thermo- and pH-responsive polymers in drug delivery, *Adv. Drug Deliv. Rev.*, **58** (2006) 1655–1670.
50. M. Heskins, J. E. Guillet, Solution properties of poly(N-isopropylacrylamide), *J. Macromol. Sci. Chem.*, **A2** (1968) 1441–1455.
51. Y. H. Bae, T. Okano, S. W. Kim, Temperature dependence of swelling of cross-linked poy(N,N'-alkyl Substituted acrylamides) in water, *J. Polym. Sci. Polym. Phys.*, **28** (1990) 923–936.
52. T. L. Lebedeva, O. I. Mal'chugova, L. I. Valuev, N. A. Plate, FT-IR spectroscopic investigation of the hydrophilic-hydrophobic balance in aqueous solutions of N-alkyl substituted polyacrylamides, *Polym. Sci.*, **34** (1992) 794–798.

53. X. Sun, Y. Huang, D. E. Nikles, FePt and CoPt magnetic nanoparticles film for future high density data storage media, *Int. J. Nanotechnol.*, **1–3** (2003) 1–19.
54. H. G. Bagaria, E. Ada, M. Shamsuzzoha, D.E. Nikles, D.T. Johnson, Understanding mercapto ligand exchange on FePt nanoparticles, *Langmuir*, **22** (18) (2006) 7732–7737.
55. S. Sun, H. Zeng, D. B. Robinson, S. Raoux, P. M. Rice, S. X. Wang, G. Li, Monodisperse MFe_2O_4 (M = Fe, Co, Mn) nanoparticles, *J. Am. Chem. Soc.*, **126** (2004) 273–279.
56. D.-H. Kim, D. E. Nikles, D. T. Johnson, C. S. Brazel, Heat generation of aqueously dispersed $CoFe_2O_4$ nanoparticles as heating agents for magnetically activated drug delivery and hyperthermia, *J. Magn. Magn. Mater.*, **320** (2008) 2390–2396.
57. B. S. Fang, Y.-Y. Pei, M.-H. Hong, J. Wu, H.-Z. Chen, In vivo tumor targeting of tumor necrosis factor-[alpha]-loaded stealth nanoparticles: Effect of MePEG molecular weight and particle size, *Eur. J. Pharm. Sci.*, **27** (2006) 27–36.
58. I. Ankareddi, M. L. Hampel, M. K. Sewell, D.-H. Kim, C. S. Brazel, Temperature controlled grafted polymer network incorporated with magnetic nanoparticles to control drug release induced by an external magnetothermal trigger, *NSTI Nanotechnol.*, **2** (2007) 431–434.
59. R. W. Baker, *Controlled Release of Biologically Active Agents*. New York, NY: John Wiley & Sons, 1987.
60. M. Gonzales, K. M. Krishnan, Synthesis of magnetoliposomes with monodisperse iron oxide nanocrystal cores for hyperthermia, *J. Magn. Magn. Mater.*, **293** (2005) 265–270.

8

Magnetic Cell Patterning

Thomas Crawford

CONTENTS

8.1 Overview..175
8.2 Background..177
8.3 Part 1: Millimeter-Scale Cellular Patterning and
    Magnetic-Field-Directed Self-Assembly for Bioengineering..............178
8.4 Part 2: Micrometer-Scale Assembly Using Magnetic Field
    Gradients for Subcellular Biology and Materials.................................186
8.5 Part 3: Nanometer Scale Magnetic-Field-Directed Self-Assembly
    for Optics, Electronics, and Bioengineering 193
8.6 Conclusion: Outlook for Science and Future Applications
    Using Magnetic Nanoparticles in Extreme Field
    Gradients..198
References..200

8.1 Overview

Francis Bitter first detailed his work observing magnetic inhomogeneities, that is, domain boundaries (Figure 8.1), using a fluid suspension of maghemite (Fe_2O_3) particles in 1931.[1] Hereafter known as the "Bitter Technique," the use of iron filings, micro- and nanoparticle ferrofluids, and other materials have long been staples of physical science demonstrations from elementary school through college. Broadly, the reason magnetic particles are attracted to these inhomogeneities is because a spatially changing magnetic field is present, that is, a magnetic field gradient. The orientation energy of a magnetic particle in a magnetic field is given by

$$E = -\mu_0 \vec{m} \cdot \vec{H} \tag{8.1}$$

in SI units, with magnetic moment m in Am^2, magnetic field H in A/m, and the permeability of free space $\mu_0 = 4\pi \times 10^{-7} N/A^2$ yielding E in Joules. Force is the gradient of this energy, and after applying appropriate vector

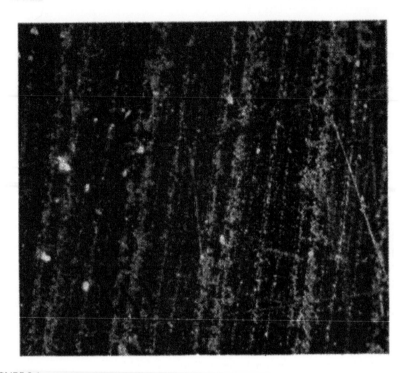

FIGURE 8.1
Pattern obtained on a sample of nickel. Magnification ×47. On the original photograph some of the lines are distinctly seen to be double, while others are single. (From Bitter, F., *Phys. Rev.*, 38, 1903–1905, 1931.)

identities for the magnetic field,[2] one can write the force on a particle with magnetic moment m as

$$\vec{F} = \mu_0 (\vec{m} \cdot \nabla) \vec{H} \qquad (8.2)$$

Note this force arises from spatial nonuniformity in the magnetic field that attracts the magnetic particle. In such cases, the field decreases with distance away from a localized field source, and thus the field gradient vector is directed toward the field source, opposite to the field itself. One can picture the process as follows. The local magnetic field magnetizes the particle—paramagnetic, ferromagnetic, superparamagnetic, and so on—and then the magnetized particle is attracted toward the direction of increasing gradient, which is opposite to the direction of the aligned moment.

This magnetic force—which is why magnets stick to refrigerators, giant electromagnets can pick up entire automobiles, and entire trains can be levitated—allows one to assemble magnetized objects into structures. In addition to these macroscopic manifestations of this force, it also works at the nanoscale. While the concept of self-assembly is well established, as discussed in recent review articles,[3,4] magnetic field-directed self-assembly has recently become a separate subfield within self-assembly.[5] Combined with the recent interest in magnetic nanoparticles,[6,7] magnetic field gradient-driven assembly has seen strong interest over the past decade in multiple fields.

This chapter is organized as follows. First, the history of magnetic field gradient assembly is contextualized by discussing magnetic filtration and chaining of magnetic particles. Next, recent studies in the fields of bio and tissue engineering utilizing micrometer- to millimeter-scale magnetic field gradients to pattern cells, control cell behavior, and regulate cellular uptake are reviewed. The use of micrometer-scale gradients to assemble nano- and microscale particles into complex assemblies and composites is discussed next. Finally, the use of nanoscale field gradients to assemble nanoparticles into hierarchical macroscale structures is reviewed. Here, changes in gradient strength relative to nanoparticle diameter becomes important for understanding how to control the assembly process.

Magnetic field gradient-directed assembly acts at length scales spanning at least 7 orders of magnitude, from nanometer to centimeter, and likely beyond. If this technique can be generalized across multiple material archetypes, at these length scales, it could become the dominant assembly technique for hierarchical construction of macroscopic materials for everything from electronics and optics, to biomaterials and tissue engineering. The ability to control and enhance novel properties assigned to nanoscale "bricks" across such a range of length scales is key to realizing an entirely new paradigm for materials manufacturing.

8.2 Background

In the 1970s, interest arose in using magnetized materials to deliberately attract magnetic particles for the purpose of filtering or materials separation.[8] The intent was to design a better sewage treatment technology in which Fe_2O_3 particles would be added to activated sludge and become nucleation sites for attracting flocs of sludge. These flocs would then be captured by a mesh of ferromagnetic wires, that is, steel wool. Later, this class of filters came to be known as high-gradient magnetic separators (HGMS), as interest continued in using ferromagnetic wires in an external field as local sources of field gradient for the purpose of attracting paramagnetic materials to the wire surface.[9–12] For example, 125-μm Ni wires were used for investigating this filtration technique with 1–20-μm diameter paramagnetic particles such as $Mn_2P_2O_7$.[9] These techniques were later applied for submicron particles, with a key difference being the need to include Brownian motion and diffusive particle transport in the simulations of particle capture.[10–12] These

HGMS studies also lead to a figure of merit for the particle diameter, d, at which Newtonian particle transport ceases to dominate and diffusive transport must be considered,[10]

$$|F|d \leq k_B T \tag{8.3}$$

where k_B is Boltzmann's constant and T is temperature. This figure of merit compares the work done by a constant force in moving the particle a distance of one diameter to the thermal energy. Gerber then shows in a table that for magnetite (Fe_3O_4) particles, diffusive transport dominates for nanoparticles with diameters less than 40 nm.[10] Importantly, this calculation was performed for a collector wire with a diameter of 100 μm, that is, a limited range of forces with gradients that can be assumed constant at nanoscale lengths.

Along a similar time frame, studies of colloidal magnetic fluids demonstrated that magnetic particles tended to form chains in the presence of an applied magnetic field.[13–15] Here, a big challenge was to develop sufficiently capable models to predict chain-like agglomeration of colloidal magnetic nanoparticles in an external field.[16–18] There has been continuous work on nanoparticle chaining since that time, including recent efforts to understand how chaining can affect measured response in magnetic resonance imaging (MRI) and magnetic fluid hyperthermia among other biomedical applications.[19,20]

8.3 Part 1: Millimeter-Scale Cellular Patterning and Magnetic-Field-Directed Self-Assembly for Bioengineering

Magnetic nanoparticles have seen tremendous growth in research aimed at possible biological and bioengineering applications.[21–23] From the original effort to separate cells with magnetic field gradients,[24] now real tissues can be grown from cells 10–100 μm in diameter into features that are millimeter scale.[25] Thus, we begin the journey across 7 spatial orders of magnitude by discussing several recent efforts to use magnetic force gradients to assemble biological materials at millimeter-length scales.

One such example is using spatially patterned magnetic field gradients to create patterns of cells.[26–28] Figure 8.2 shows a recent example of using magnetic gradient forces to create patterns of three-dimensional (3D) endothelial cell spheroids by loading the individual cells with superparamagnetic Fe_3O_4 nanopaticles[26] (SPION is an acronym commonly found in the biomagnetism literature, standing for superparamagnetic iron-oxide nanoparticle). Whatley et al. incubated endothelial cells with silica-coated Fe_3O_4 nanoparticles and, after performing cytotoxicity studies, grew them into cell spheroids using a robotic spheroid maker of their own design. After 3 days of spheroid

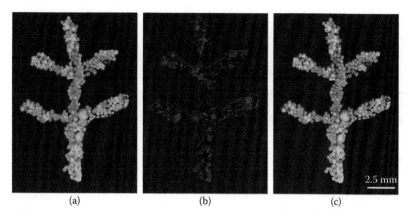

(a)   (b)   (c)

FIGURE 8.2
Image of endothelial cell spheroids patterned onto a branched magnetic template cut from a magnetic sheet. (a) is stained for Actin, while (b) is stained for cell nuclei. (c) is the combined image. Here the template was removed after 48 hours, while the images were taken at 10 days in culture. This result shows that the cell spheroids grow and merge, and importantly, maintain their pattern even after the magnetic template is removed. (From Whatley, B. R. et al., *J. Biomed. Mater. Res.*, 2013.)

growth and removal of the spheroids from the agarose microwells used for fabrication, they were monitored with fluorescent staining to determine whether the nanoparticle-loaded cells in the spheroid remained viable.

They designed branched magnetic templates using CAD which were then cut from Nd magnetic sheets with a computerized cutting tool. Tissue culture wells were placed on top of these patterned templates and the cell spheroids were incubated on a shaker to help bring the spheroids near the magnetic fields at the bottom of the wells. After 48 hours, the magnetic template was removed from below the wells, and the patterns were imaged with confocal microscopy (Figure 8.2). Even after 10 days, the cellular patterns closely matched the pattern of the original magnetic field template. These patterns were maintained even after being extensively washed and shaken, meaning the cells had fused together in the pattern dictated by the magnetic field, and that the pattern was not degrading with time. Importantly, the branched structure suggests tissue-like patterns can be used as template for cell growth according to user design. Finally, the authors point out that by changing the properties of the nanoparticles loaded into particular cells, heterogeneous structures with different kinds of cells can be assembled.[26] This ability to vary the size, magnetic material, and hence the magnetic force, makes it possible to create different kinds of assembled materials within a single structure, simply by adjusting the properties of the assembling magnets.

Another approach is to position magnets in patterned arrays while the cell spheroids themselves are cultured. In a study by Bratt-Leal et al., the actual location of the cell spheroid was controlled by spatially arranging magnets on the lid of the culture dish.[29] As shown in Figure 8.3, a variety of

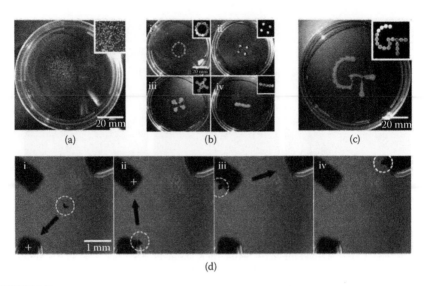

FIGURE 8.3
Image of stem cell spheroids that were aggregated with magnetic micro particles during spheroid creation. Thus, the particles are in the extracellular space. (a) Spheroid aggregation without magnetic field, (b) and (c) control of spheroid pattern through placement of magnets on lid of culture dish, and (d) spatial control of spheroid placement with a patterned array of iron magnets in PDMS. (From Bratt-Leal, A. M. et al., *Integr. Biol.*, 3, 1224, 2011.)

cell spheroid patterns could be obtained by placing the assembly magnets in different spatial arrangements. Here, murine embryonic stem cells are first aggregated into embryoid bodies (i.e., 3D cell spheroids) using forced aggregation, and then polystyrene microparticles (~4 μm diameter) are added to load the extracellular space of the spheroids with the magnetic microparticles prior to patterning. This loading of the microparticles was performed either by centrifugation or by using a large magnet below the plate (10 lb. pull strength) to force the microparticles into the spheroids.

Large aggregates were then removed from the wells and transferred to a suspension culture on a rotary shaker. As a control, the spheroids simply clustered at the center of the dish in the absence of patterning magnets (Figure 8.3a). However, by placing magnets on the lid of the dish, the authors were able to create controlled patterns of spheroids (Figure 8.3b and c). Further, embedding iron pillars in polydimethylsiloxane (PDMS) structures where the suspension was placed between the pillars, individual spheroids could be manipulated between the pillars, suggested the ability to direct individual spheroids to particular places in a tissue construct (Figure 8.3d, i through iv).[29] Importantly, the authors emphasize that because the magnetic particles are in the extracellular space, there is no affect of the microparticles on signaling or intracellular machinery. Again, as above, here the authors emphasize the ability to control heterogeneity in tissue structures. By varying magnetic field strength, pattern, and particle magnetic properties, *hierarchical* control can be exerted over tissue architecture at multiple length scales. Finally, the authors point out that by attaching morphogenic factors to the magnetic particles, locally patterned environmental signals can be delivered to particular cell spheroids and further control production of tissue from stem cells.

Sharp magnetic tips or tweezers can exert forces on cells for a variety of mechanical analyses.[30,31] Frasca et al. used ~0.5–1 millimeter-scale iron tips magnetized by a permanent magnet to attract human endothelial progenitor cells labeled with citrate-coated Fe_2O_3 nanoparticles.[31] In this study, the magnetized tips could attract cells to form 3D assemblies (Figure 8.4). By focusing the magnetic field at a local spot, that is, tuning the extent and strength of the magnetic field gradient, as well as by controlling the number of cells, Frasca et al. were able to understand the details of the 3D grouping of cells that were created. The cells took in the magnetic nanoparticles via endocytosis, similar to that in Whatley et al.'s study, and the team performed single-cell magnetophoresis in 17 T/m field gradients, achieving cell velocities from 20–110 μm/s after a 2-hour incubation. The cellular iron load was controlled by varying the length of time the cells were incubated with the nanoparticles.

For the 1-mm iron tips, field gradients of ~1000 T/m could be applied at 0.5 mm from the tip, and over these length scales, the gradient was non-uniform. For a cylindrical tip, the gradient was more uniform over a larger distance, while a shaped tip produced a larger gradient that decreased more quickly with distance. For a cell labeled with 12 pg of iron, the authors estimated a 900 pN force could be applied to the cell by the field gradient, large enough to overcome Brownian motion and any other forces that might be present in a cellular suspension. The degree to which the cells were packed in the 3D assembly depended on the total magnetic load in the cell, with lower magnetic loads leading to more loosely packed assemblies. Interestingly, cell–cell interactions appeared not to impact assembly during the first 10 minutes, being dominated by the magnetic assembly force. Finally, the authors noted that the efficiency of aggregating cells depended on both the

FIGURE 8.4
Magnetic tips can assemble cells into 3D constructs (top 3 panels), while without a magnetic field, the cells spread horizontally. (From Frasca, G. et al., *Langmuir*, 25, 2348–2354, 2009.)

cell's total magnetic moment and the size of the magnetic field gradient, as expected by Equation 8.2 above.

In contrast to forming multicellular assemblies by using magnetophoresis to attract cells loaded with magnetic nanoparticles, it is also possible to use so-called *negative* magnetophoresis to assemble cells.[32] Here, the cells are not loaded with magnetic particles, but rather are loaded in ferrofluid, and thus, it is the *absence* of magnetic materials inside the cells that allows the cells to be transported through a uniformly magnetized fluid. Figure 8.5 shows the formation of chains of endothelial cells stained with dye and imaged with confocal microscopy. Here, the magnetic particles are coated with bovine serum albumin (BSA) and suspended in endothelial growth medium. Linear chains of cells were formed by applying a 100 Oe magnetic field to slides containing the BSA solution and the cells. Importantly, the cell chains remained intact after removal of the magnetic field.

The average chain length grew with the length of time the magnetic field was applied and was observed to be consistent with diffusion-limited aggregation, that is, a power law dependence for chain length (Figure 8.5). Since the tissue growth suspension for this case of negative magnetophoresis has to also be a ferrofluid, the authors performed cell viability studies that showed >95% cell viability after exposure to the ferrofluid. Even after 2 hours of ferrofluid exposure, the cells were still able to proliferate normally in a tissue culture. Finally, the authors demonstrated that this patterning process works in cell culture medium. The assembled cells could form single- or multiple-cell width chains depending on the nanoparticle concentration, with more concentrated ferrofluid leading to wider cellular structures. In addition, once the assembly was nucleated, the ferrofluid could be removed and replaced with growth medium without disrupting the nanoparticle chain structure.

In addition to creating arrays of cells and cell spheroids, single cells can be patterned and manipulated with magnetic fields.[33,34] Ino et al. demonstrated the ability to pattern single cells into square arrays (Figure 8.6).[34] They used magnetite cationic liposomes (MCLs), which are liposomes with embedded Fe_3O_4 nanoparticles. These MCLs were designed for uptake into target cells through electrostatic attraction to the cell membrane. Thus, cells labeled with MCLs can easily be manipulated with magnetic field gradients. In their work, they created an iron "pin" that had 6000 100 × 100 × 300 μm^3 pillars of machined using wire-EDM into its surface. This pin was then used for magnetic force-based tissue engineering, that is, to assemble arrays of cells on the pillars. The cells used in this study were mouse NIH/3T3 fibroblasts, which were cultured with MCLs for 4 hours to yield 3T3 cells labeled with magnetic nanoparticles. The pin was then placed on a cylindrical permanent magnet and located beneath a tissue culture dish made from Silicon. Different thicknesses were used to vary the magnetic force applied to assemble the 3T3 cells.

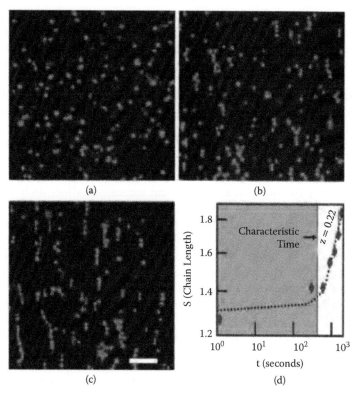

(a) (b)

(c) (d)

FIGURE 8.5
(a-c) Negative magnetophoresis for assembling linear chains of cells in a ferrofluid consisting of BSA-coated nanoparticles. The average chain length grows linearly in time, consistent with a power law diffusion limited aggregation mechanism (d). (From Krebs, M. D. et al., *Nano Lett.*, 9, 1812–1817, 2009.)

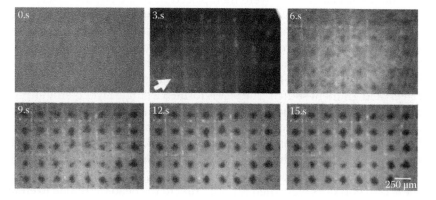

FIGURE 8.6
Fibroblast cells assembled onto wire-EDM machined pillars. Images show a time series of labelled cells assembling onto the nanostructured mesas. (From Ino, K. et al., *lab chip*, 8, 134, 2007.)

Ino et al. observed that cells were patterned into arrays in a mere 15 seconds and Figure 8.6 shows a series of photographs from 0 to 16 seconds that demonstrated the speed of patterning the 3T3 cells onto the iron pin features. By reducing the initial cell concentration, single cells could be assembled on 48% of the iron pillars. The authors then treated the assembled cells with a DNA-damaging agent, mitomycin C (MMC), and demonstrated that the MMC prevented 31% of the cells from changing shape compared with the 14% that stayed the same without MMC. Thus, the patterned cell array can be used to detect the action of a chemical agent and obtain cell-by-cell statistics.

Since endothelial cells are dependent on how they are adhered to a substrate, how they distort in reaction to local forces can affect their viability, that is, can lead to apoptosis. Polte et al. structured the surface of a patterned magnet array to not only assemble cells into patterns but also control their fate.[35] The idea was to create a controlled adhesion substrate, where magnetic field gradients applied adhesive forces by attracting the magnetic beads bound to the endothelial cells. Polte et al. took a unique approach to creating locally high-magnetic field gradients. They used a femtosecond laser to etch the surfaces of silicon and steel and form nanoscale spikes. In the silicon case, these spikes are coated with thin-film Co to create a continuous surface of localized magnetic field gradients. By replicating these textured surfaces in a macroscopic cellular-sized pattern, they create a multiplexed array of locally controlled adhesion sites. The reason for replacing Si with steel was to increase the robustness of the materials and simplify the fabrication process. Here, they etched 50×50 μm^2 mesas in a patterned array and then created the nanospikes on the surfaces of the mesa regions. The nanospikes were ~300 nm wide, 700 nm high, and were spaced 400 nm apart on average.

After coating the patterned steel array with a thin layer of PDMS, they were treated with F-127 to make the surface nonadhesive to cells, proteins, or the magnetic beads. They were then placed in tissue culture dishes such that permanent magnets could be placed below the field gradient source to magnetize the steel. To demonstrate that this system works, micrometer-sized magnetic particles treated with RGD cell-binding peptide were shown to settle on the PDMS surface in random order without the external magnetic field, but to take on the mesa pattern when the magnetic field was applied to the steel field gradient source. The beads formed a single layer on the surface and were able to withstand gentle agitation of the medium as long as the field gradients were applied. Force measurements showed that at the surface of the PDMS film, 5 μm from the steel surface, 0.9–1.3 nN could be applied to a single magnetic bead, which is on the order of the force living cells apply to local adhesions.

These concentrators were then exposed first to the beads and then to endothelial cells at low 40 cells/mm^2 densities to prevent more than one cell from adhering to a given island. Figure 8.7a shows fluorescence from the beads

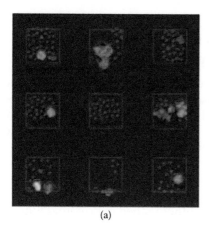
(a)

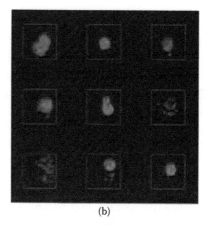
(b)

FIGURE 8.7

Magnetic microbead and endothelial cell adhesion to a patterned array of stainless steel mesas with nanostructured surfaces to provide local magnetic field gradients. (a) shows single-cell adhesion on the bead/mesa surface when the cells are added to the culture dish after the magnetic particles, while (b) shows the situation when the cells and beads are adhered prior to being added to the culture dish. (From Polte, T. R. et al., *Biomaterials*, 28, 2783–2790, 2007.)

and cells themselves, demonstrating that the magnetic force on the bead together with the RGD adhesion can hold a cell in place over the patterned magnetic array. Figure 8.7b shows cells prebound with RGD-coated beads in suspension can also be assembled onto the square islands. As long as the magnetic field was applied, the cells remained in place. Note that the areas of the cells spread more as the strength of the underlying magnetic field was increased (here by adding more permanent magnets below the tissue culture dish). If the magnets were then removed, all the cells detached themselves from the PDMS substrate. For these endothelial cells, detachment from this controlled-adhesion substrate means apoptosis. Thus, the magnetic field itself can act like a switch, in this case to trigger cell death. The authors point out that such a switch could be used to recharge the array with new live cells, for example in a biosensor.

The studies discussed in this section clearly demonstrate the potential of magnetic field gradient forces to assemble biomaterials at the cellular to multicellular level. Many of these studies, specifically Polte et al., highlight that magnetic field gradient forces can act over spatial lengths from 500 nm (Polte's nanospikes) to the millimeter–centimeter scale of the external magnets, which were added to increase the cell adhesion onto these 50-µm spaced arrays. Similar to the biomaterial studies with larger scale magnets, interest in assembly of structures over the past decade has led to novel work at the micrometer scale, using magnetic forces to assemble and manipulate magnetic particles for a variety of potential biological and sensing applications.

8.4 Part 2: Micrometer-Scale Assembly Using Magnetic Field Gradients for Subcellular Biology and Materials

The ability to pattern magnets into structures using microscale photolithographic techniques has enabled an exciting body of research on magnetophoresis and magnetic field-directed self-assembly.[4,5] Micrometer- and submicrometer-patterned magnets have been used to assemble micro- and nanoparticles into a variety of structures.[36–42]

Electromagnets (i.e., micropatterned current-carrying wires), permanent magnet materials, and soft ferromagnets magnetized by a large-scale external fields have all been used for directing self-assembly of magnetic particles at the microscale. Figure 8.8 shows a lithographed microtoroid without current (Figure 8.8a) and with a 350 mA current (Figure 8.8b). 10–20-nm diameter Fe_3O_4 nanoparticles are encapsulated in polystyrene/carboxylic acid having a total diameter of 1–2 μm, and are shown in Figure 8.8b as assembled by the magnetic field gradient, which effectively traps the microparticles in the center of the toroid.[36] These toroids were later used to trap magnetotactic bacteria by trapping chained magnetic nanoparticles located within the bacteria.[43]

In addition to electromagnets, thin-film islands of Co have been patterned into arrays for trapping paramagnetic beads.[41] The Co islands are then magnetized using an external permanent magnet and micrometer-scale superparamagnetic beads were assembled onto the array. Depending on the conditions, beads assemble both in the center and the edges of the Co islands. Chaining of particles between islands was also observed when

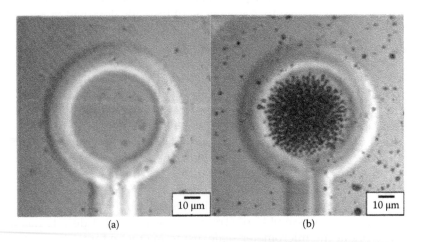

(a)   (b)

FIGURE 8.8

Micrometer-scale current-carrying toroidal wire for trapping magnetic nanoparticles. (a) Ring with no current and (b) 350 mA current through the wire assembles 1–2 μm diameter polystyrene particles with ~10–20 nm magnetite cores. (From Lee, C. et al., *Appl. Phys. Lett.*, 79, 3308–3310, 2001.)

the external field is aligned with the field of the islands. However, when the external field is opposite to the island field, the beads are pulled to the center of the islands.

Even more exotic assemblies are possible. Figure 8.9 shows bundles of rod magnets. Consisting of alternating regions of Ni and Au, electrodeposited

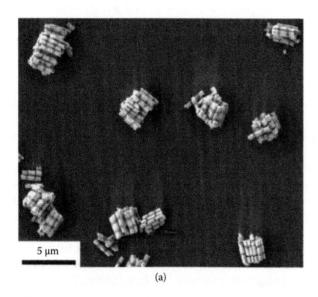

5 µm

(a)

500 µm

(b)

FIGURE 8.9
Rods consisting of alternating Ni (gray) and gold (lighter) sections are bundled together via self-assembly. (From Love, J. et al., *J Am Chem Soc.*, 125, 12696–12697, 2003.)

into porous alumina filters, these rods were first sonicated and then dropped onto Si wafers, where they self-assemble into the bundles shown in the Figure 8.9b.[42] In Figure 8.10, magnetic microwells can be turned on and off with an external field to trap and release superparamagnetic beads from within a well.[44] Many similar studies explored how nanoparticles and beads could be assembled parallel to external fields, along domain walls, and on various types of magnetically patterned surfaces at the microscale.[45–48]

To better understand magnetic field-driven assembly at the microscale, a number of studies modeled magnetic properties of colloidal fluids in the presence of heterogeneous, patterned magnetic fields.[40,49–51] Figure 8.11 shows a calculation by the Furlani group that modeled nanoparticle trajectories as moving particles entered a region of patterned magnetic field gradients. As the entry height of the particles is varied, the ability of the gradients to perturb and or capture the particle changes dramatically. Interestingly, for a number of initial heights, the particle is either captured by the first patterned magnet or the last magnet, with only a few trajectories leading to particles captured in the magnets in the center of the array. This interesting calculation highlights not only the complexity of these magnetically driven fluid systems, but also the possibilities for engineered control over how precisely the particles move above a patterned array of magnetic field gradients.[49]

With basic efforts to control nanoparticle patterns demonstrated, more complex structures and composites were assembled by adding additional ability to vary the external field, as well as by adding multiple sizes of particles with varying magnetic moment.[52–55] In addition, the concept of negative

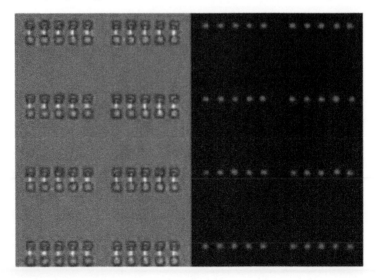

FIGURE 8.10

Assembly of superparamagnetic microbeads onto patterned micromagnets. By reversing the direction of a perpendicular bias field, the beads can be directed to two different locations (one end of the micromagnet or the other).

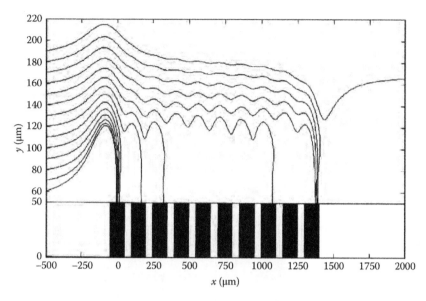

FIGURE 8.11

Calculation of trajectories for micro/nanoparticles entering a fluid cell with a series of micro-patterned magnets. As the height is varied, the capture efficiency changes. (From Furlani, E., *J. Appl. Phys.*, 99, 024912, 2006.)

magnetophoresis, where a nonmagnetic particle can be transported through a magnetized fluid, was also demonstrated.[56–59] Figure 8.12 shows an example of multicomponent particle assembly by the Yellen group (Erb et al.[59]). By adding paramagnetic and nonmagnetic particles to a ferrofluid, where the nonmagnetic particles behave like a diamagnet in a magnetized ferrofluid, they demonstrated arrangements of particles that resemble electrostatic charge configurations. Structures include axial and linear quadrupoles, octopoles, and other multipole symmetries.

In addition to using both paramagnetic and nonmagnetic particles, by varying the size of the particles, more complex structures are possible. Figure 8.12a (left) shows ferrofluid plus two different paramagnetic particle sizes. Here the smaller particles assemble at the poles of the larger particles. Figure 8.12a (right) shows ferrofluid plus differently sized paramagnetic and nonmagnetic particles. In this case, the nonmagnetic particles align anti-ferromagnetically with respect to the paramagnetic particles, and thus the smaller nonmagnetic particles assemble around the equator of the larger paramagnetic particles. Figure 8.12b (left) shows 4-component suspensions, where small- and medium-sized nonmagnetic particles are combined with larger paramagnetic particles. At the right concentration, the smaller non-magnetic particles can take on the same magnetization as the large para-magnetic particles, while the medium-sized nonmagnetic particles behave diamagnetically and align opposite to the paramagnetic particle, that is, along

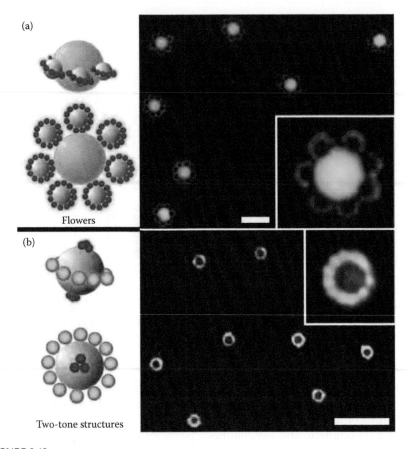

FIGURE 8.12
Hierarchical multicomponent self-assembly in ferrofluid. By using both nonmagnetic and paramagnetic particles having different sizes and concentrations, a rich tapestry of assembled superstructures can be created. (From Erb, R. M. et al., *Nature*, 457, 999–1002, 2009.)

the equator. Figure 8.12b (right) shows the case of small and large nonmagnetic particles assembling with a medium-sized paramagnetic particle. In this case, the large particles are diamagnetic with respect to the ferrofluid and the medium paramagnetic particles (which are along the equator). The smaller nonmagnetic particles then align opposite to the paramagnetic particles. All of these configurations depend on the concentration and magnetization of the ferrofluid and the paramagnetic species, as well as the external field. In their study, Erb et al. presented phase diagrams that showed applied field versus volume fraction and could identify where poles, rings, and random formations occurred. It is near the phase boundaries where the complex, fractal flower-like formations occur. In a related work, Li and Yellen[60] showed that the ring structures could be tuned by adjusting the ferrofluid volume fraction and the magnitude of the external field. 10-µm nonmagnetic beads were combined with 2.7-μm superparamagnetic beads as shown in Figure 8.13, which has the smaller paramagnetic particles assembling into equatorial rings on the outside of the nonmagnetic particle, similar to that shown in Figure 8.12. By adjusting the concentration of the ferrofluid, the number of particles along the equator can be tuned one particle at a time, from 2 to 3 to 4 and all the way to 9 particles. Again, these results suggest a rich variety of phenomena that can be explored at the microscale.

Using (nearly) equal-sized magnetic and nonmagnetic (diamagnetic) beads more than 20 different structures can be self-assembled in different concentrations of ferrofluid.[61] In addition to the square lattice of equal amounts, honeycomb and Kagome lattices, a variety of stripe phases, domains, other types of lattices, and many varieties of combination chains were assembled using this technique. These different phases were all the result of potential energy minimization using the ferrofluid, and the relative concentrations of the two types of beads.

Similar kinds of structures have been assembled using patterned magnetic features instead of ferrofluid. By applying currents to magnetize these features; for example, by moving domain walls, magnetic beads can be trapped and manipulated dynamically.[62-67] One recent example combines assembly with patterned micromagnets[47] with the multicomponent colloidal assembly,[61] to yield even more interesting colloidal microstructures.[68] Here, paramagnetic particles assemble on patterned Ni magnets, while nonmagnetic (diamagnetic) species assemble onto the voids between the patterned magnets. By combining ionic concentration gradients in the colloidal solutions with the patterned magnetic field gradients, live bacteria, nanoparticles, and ions can be positioned onto these patterned magnetic arrays (Figure 8.14).

As shown above, self-assembly of magnetic particles using magnetic-field gradients at the microscale has exhibited a rich variety of complex

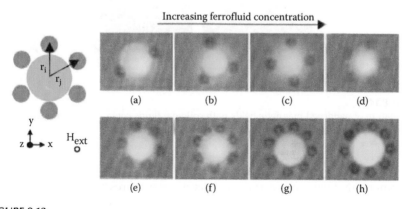

FIGURE 8.13
Two-component ring assembly tuning. By varying the ferrofluid volume fraction the number of particles in the ring structure can be tuned one particle at a time. (From Li, K. H.; Yellen, B. B., *Appl. Phys. Lett.*, 97, 083105–083105–3, 2010.)

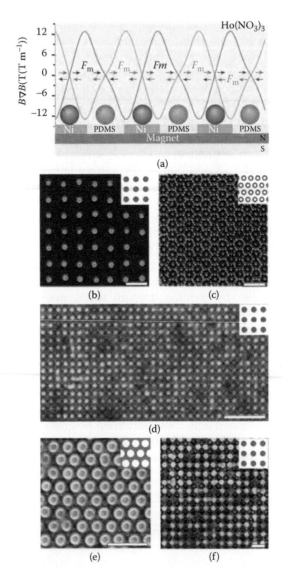

FIGURE 8.14
Combination of patterned magnetic fields with colloidal fluids containing paramagnetic and diamagnetic particle species to achieve more complex 2D colloidal assembly. (From Demirörs, A. F. et al., *Nature*, 503, 99–103, 2013.)

phenomena that can even mimic real crystal structures.[61] And these assembly techniques are being extended to a variety of nonmagnetic species, where magnetism is used for assembly, but other interesting materials are assembled.[68] Moreover, these techniques are being extended to modify the behavior of composite materials.[69] Platelets coated with small concentrations of magnetic nanoparticles (0.5% by volume) can be aligned inside a composite material. Importantly, the phase diagram for magnetic field versus platelet size identifies a region where an extremely weak magnetic field, merely 10x the earth's field (~5 Oe), can align these platelets inside a composite. Thus, small quantities of superparamagnetic particles and weak magnetic fields can exert a large impact on the ultimate properties of a composite material.

One reason for such a rich exploitation of phenomena at the microscale is possibly the easy visualization of these colloidal systems using optical and fluorescence microscopy with readily available paramagnetic and non-magnetic beads prefunctionalized with libraries of fluorophores. Moreover, patterning magnets at the microscale is significantly easier than creating patterned magnets at smaller length scales. However, as will be seen in Section 8.5 (Part 3), these same kinds of self-assembly techniques can be employed at the nanoscale.

8.5 Part 3: Nanometer Scale Magnetic-Field-Directed Self-Assembly for Optics, Electronics, and Bioengineering

One of the key features of using more localized magnetized field sources is they offer larger field gradients, and hence larger forces, which enables one to create assemblies using smaller magnetic particles, for example, approaching 10 nm diameter. Using the figure of merit from the HGMS literature, presented in Equation 8.3 above, one way to avoid Brownian diffusion randomizing the assembly of nanoparticles is to increase the force commensurate with a decrease in particle size. For the magnetic force to perform work comparable to thermal energy over a 10 nm distance, (decreasing particle diameter from 40 nm to 10 nm), the magnetic force must increase by a factor of 4. However, since the particle moment is proportional to d^3 the force actually decreases by a factor of 64 compared to the force on a 40-nm particle. Therefore the work done on a 10-nm particle to move it 10 nm decreases by a factor of 256 for the same field gradient. To compensate the magnetic field gradient must be increase by more than **250x** in order to direct assembly of smaller nanoparticles with minimal thermal effects! As will be seen below, such larger gradients are possible using nanoscale patterned magnetic field sources.

Traditional magnetophoresis experiments are done in field gradients that are less than 1000 T/m.[67] The millimeter-scale experiments described above are done well below this limit, using macroscopic permanent magnets (i.e., 0.3 T/cm ~30 T/m for a 3 kOe NdFeB permanent magnet). For the micrometer-scale work, the gradients approach this upper boundary (i.e., $0.0027T/10 \, \mu m = 270$ T/m).[36] Note these are just rough estimates.

Much larger gradients can be achieved by decreasing the size of the magnetized region that emits a field. In work by the Whitesides group from 2003, 80-nm diameter Co–Ni rods spaced by varying thicknesses of Au were used to assemble 8-nm diameter Fe_2O_3 particles from solution.[70] Similar sized gradients can be generated by magnetic nanoparticles themselves, assuming they behave like dipoles. In a uniform magnetic field, this gradient is what causes nanoparticles to chain into macroscopic assemblies.[43,71] Figure 8.15 shows Korth et al.'s chains of 25-nm nanoparticles assembled in 100 mT fields after being cast from a Toluene dispersion.[71] Similar chains were observed by Lee et al. after trapping magnetotactic bacteria with micrometer-scale patterned electromagnets. The bacteria were lysed in place and the cellular membranes washed away to leave behind nanoscale chains of magnetic nanoparticles, where both linear and closed chain structures were observed.[43]

In recent simulations, the Furlani group predicted that by using a core–shell structure, where the core is magnetic and the shell could be silica, the diffusion figure of merit can be overcome by controlling the magnetic force via the size of the core, in their case a 30-nm diameter core with a 15-nm thick silica shell.[72] Moreover, by using confined magnetic ring elements ~1000 nm in diameter but with 100 nm thickness, together with external fields, they predict assembly of near perfect rings, square frames, and cross structures (Figure 8.16) built from evenly spaced 60-nm diameter particles (with 30-nm diameter magnetic cores). Key to building such predicted assemblies are patterned magnets with dimensions ~100 nm., Urbach et al. used electrodeposition into nanotemplates to create ~100-nm cylindrical magnets with

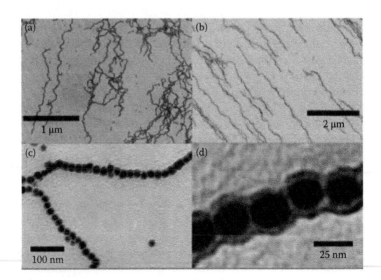

FIGURE 8.15
The 25-nm diameter nanoparticles assembled into single-nanoparticle wide chains using an external magnetic field. (From Korth, B. D. et al., *J. Am. Chem. Soc.*, 128, 6562–6563, 2006.)

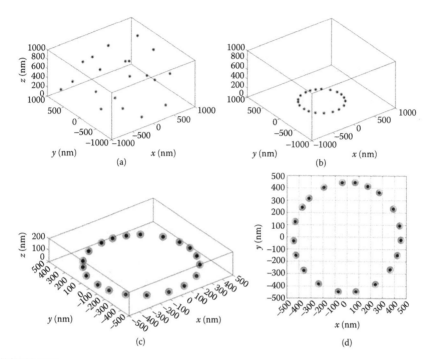

FIGURE 8.16
Simulation of magnetic-field template driven self-assembly using 60-nm diameter nanoparticles with a 30-nm magnetic core and 15-nm silica shell. Circular and square frames were also simulated. (From Xue, X.; Furlani, E. P., *Phys. Chem. Chem. Phys.*, 16, 13306, 2014.)

gradients along the length of cylinder where gaps were formed between Co-Ni layers and Au layers.

A ready source of user-patternable sub-100-nm field gradient sources does in fact exist, and it has been recently exploited by our group to create assemblies and patterns of magnetic nanoparticles at the macroscale.[73–75] This source is the disk used in a magnetic disk drive to store much of the content the world accesses on the Internet. The disk, also known as a recording medium,[76] consists of ~5-nm diameter Co grains that are deposited together with Cr that segregates the Co grains and magnetically decouples them so they act like individual nanomagnets. These 5-nm grains are contained in a continuous film of metal that is deposited onto the disk itself. The grains are then magnetized by the recording head, a 1–2 T electromagnet capable of creating regions of oppositely magnetized grains with spacings as small as 12 nm. Such small spacings lead to transition regions a mere 1–3 nm in size, and from these regions, enormous gradients are emitted, approaching 20 MT/m, 20,000 times larger than the gradients discussed above, and more than enough to minimize random thermal motion during nanoparticle assembly. Importantly, however, these gradients can be user controlled to cover the entire 3-in. diameter surface of the disk drive.

While the recorded patterns are the "bits" that make up the data stored on the disk, there is nothing to prevent recording patterns that are designed for self-assembly of nanoparticles. Figure 8.17a shows such an optical image of micrometer-scale patterns that are built by self-assembling magnetic nanoparticles onto the nanoscale gradients on a disk. Our group then spin coated it with a curable liquid polymer that when peeled from the surface transferred the nanoparticles to a standalone polymer film (Figure 8.17d). Figure 8.17b shows the polymer film and Figure 8.17c shows optical diffraction from a series of parallel lines with a 750 nm spacing that were fashioned into an all-nanoparticle diffraction grating.[75] The inset in Figure 8.17a and Figure 8.18 a–d shows schematically how this technique, which we call pattern transfer nanomanufacturing™ (PTNM), works.

In addition to building a prototype polymer diffraction grating (Figure 8.18e–g), Henderson et al. demonstrated that by combining the alternating magnetic fields of these in-plane magnetized regions with an external field, the attraction and repulsion could allow for interleaved multilayer assemblies. Figure 8.19 shows how the multilayer assembly works schematically at the top. In Figure 8.19a, the up-directed external magnetic field premagnetizes the nanoparticles such that they are attracted to the upward-directed fields from the transitions recorded into the disk medium. However, these particles are repelled from the alternate transitions because the fields are opposite in direction. After the assembly is created (and the optical image shows that for weaker fields, there is still some assembly on the alternate transitions), the polymer film was applied, the external field reversed, and the nanoparticles assemble on the tails of the arrows, interleaved with respect to the lower level (brighter regions in the right image compared with the brighter regions in the left image).

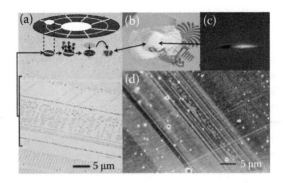

FIGURE 8.17
Experimental demonstration by author's group of nanoscale templated self-assembly of magnetic nanoparticles onto disk drive recording media using 15 MT/m field gradients. These patterns were transferred from the disk surface to a standalone polymer film. (From Henderson, J. et al., *Nanotechnology*, 23, 185304–185304, 2012.)

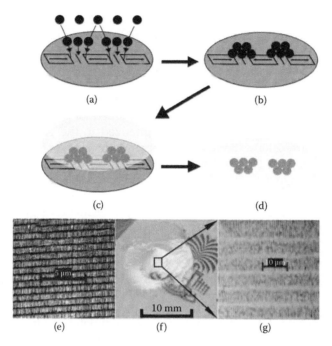

FIGURE 8.18
All-nanoparticle diffraction grating assembled on a template magnetically-patterned into a
magnetic recording disk. (From Ye, L. et al., *Opt. Express*, 21, 1066–1075, 2013.)

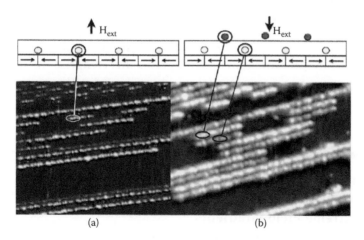

FIGURE 8.19
Interleaved bilayer templated assembly using the alternating field gradients on a recorded disk
drive together with an external field. The nanoparticles assemble first on upward field transitions, and then by reversing the external field, they assemble on the downward field transitions. (From Henderson, J. et al., *Nanotechnology*, 23, 185304–185304, 2012.)

Excitingly, this experimental demonstration that the combination of external fields and localized field gradients can both attract and repel magnetic
nanoparticles,[74] proves that a key feature of Furlani's recent simulations[72]
works in practice.

Finally, the disk drive industry dramatically modified its recording media
in ~2007. The recording process to template the media used for nanoparticle assembly in Figures 8.17 through 8.19 is called longitudinal recording,
primarily because the Co grains are magnetized in the plane of the thin
film.[76] In the early 2000s, the industry began to develop an alternate media
technology to replace longitudinal media. The issue was that the recording
head used to magnetize the Co grains could not produce enough field to
keep shrinking the size of the grains. The replacement technology orients
the Co grains perpendicular to the disk surface. Known as perpendicular
recording, the recording industry placed a soft magnetic layer beneath the
Co media to boost the head field and enable smaller, higher anisotropy magnetic materials to be used for the media.[76]

This switch is a major boost for using magnetic media as a reprogrammable
template for nanoscale magnetic field-directed self-assembly. Because the
grains are magnetized perpendicular to the two-dimensional (2D) surface that is the template, arbitrary shaped 2D templates can be patterned.[77]
While the longitudinal results prove the utility of this technique, and can
compete with lithography on cost and ease of template creation, longitudinal media patterns are quasi-one-dimensional (1D), because 1D is dictated
by the width of the recording head. With perpendicular recording, arbitrary 2D patterns can be created, much like the features simulated by the
Furlani group.[72,77] Figure 8.20 shows 30-nm diameter Fe_3O_4 nanoparticles
assembled into square and triangular shapes using templates created with
perpendicular recording.[77] Panel C shows a close-up of individual 30-nm
particles and the ~5–7-nm diameter Co grains that comprise the recording
media.

8.6 Conclusion: Outlook for Science and Future Applications Using Magnetic Nanoparticles in Extreme Field Gradients

The magnetic nanoparticle is a complicated material. Consisting of a magnetic or magnetic-oxide core, it is typically presumed that the spins at the
surface are magnetically dead.[78] However, recent results suggest that there
may be a unique spin state that is present at the nanoparticle surface.[79] The
surfactant that is present on the surface has been shown to affect the dynamics of the nanoparticle magnetization.[80] Finally, measurements of anisotropy

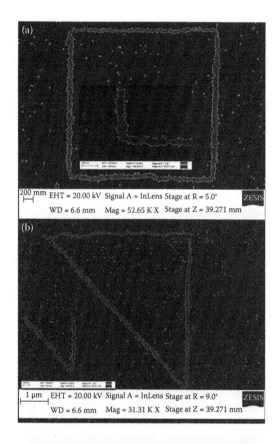

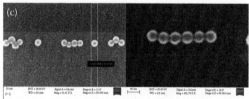

FIGURE 8.20

The 30-nm diameter Fe_3O_4 nanoparticles assembled into square and triangular shapes recorded into perpendicular magnetic recording media templates. Note the features are ~100 nm wide, that is, 3–4 nanoparticles. Panel C shows the size of the Co grains in the media together with a line of assembled 30-nm diameter particles. (Dolbashian, Pstrak, Pearson, Ye, Fellows, Mefford, and Crawford —in preparation, 2014.)

and heating of nanoparticles via alternating fields suggests the community still does not understand this unique material.[81]

The ability to use magnetic field-gradients across 7 orders of magnitude: to assemble single 30-nm particles into a pattern on a surface, into a chain in solution, and to pattern 100-mm diameter cell spheroids in a bioreactor for tissue engineering, has enormous potential to impact human health, infrastructure, and technology. Further research on assembling these materials together into larger structures, as well as on understanding how their magnetic properties change as we modify them with chemistry, could enable a tremendous variety of different applications. If sufficient varieties of materials can be built using magnetic forces to direct a magnetic core with an outer shell tuned for function, that is, for biomaterials, optics, electronics, or other applications, the ability to manufacture novel materials could be transformed in the not-so-distance future.

References

1. Bitter, F. On inhomogeneities in the magnetization of ferromagnetic materials. *Phys Rev* **1931,** *38* (10), 1903–1905.
2. Reitz, J. R., Milford, F. J., Christy, R. W. *Foundations of Electromagnetic Theory.* Reading, MA: Addison Wesley, 1960.
3. Whitesides, G. M., Grzybowski, B. Self-assembly at all scales. *Science* **2002,** *295* (5564), 2418–2421.
4. Bishop, K. J. M., Wilmer, C. E., Soh, S., Grzybowski, B.A. Nanoscale forces and their uses in self-assembly. *Small* **2009,** *5* (14), 1600–1630.
5. Tracy, J. B., Crawford, T. M. Magnetic field directed self-assembly of magnetic nanoparticles. *MRS Bull* **2013,** *38* (11), 915–920.
6. Gubin, S. *Magnetic Nanoparticles.* Weinheim: Wiley-VCH Verlag GmbH & Co. KGaA, 2009.
7. Kumar, C. *Magnetic Nanomaterials.* Weinheim: Wiley-VCH, 2009, Vol. 4.
8. Watson, J. H. P. Magnetic filtration. *J Appl Phys* **1973,** *44* (9), 4209.
9. Friedlaender, F., Takayasu, M., Rettig, J., Kentzer, C. Studies of single wire parallel stream type HGMS. *IEEE Trans Magn* **1978,** *14* (5), 404–406.
10. Gerber, R., Takayasu, M., Friedlaender, F. Generalization of HGMS theory: The capture of ultrafine particles. *IEEE Trans Magn* **1983,** *19* (5), 2115–2117.
11. Takayasu, M., Gerber, R., Friedlaender, F. Magnetic separation of sub-micron particles. *IEEE Trans Magn* **1983,** *19* (5), 2112–2114.
12. Gerber, R. Magnetic filtration of ultra-fine particles. *IEEE Trans Magn* **1984,** *20* (5), 1159–1164.
13. Martinet, A. Biréfringence et dichroïsme linéaire des ferrofluides sous champ magnétique. *Rheol Acta* **1974,** *13* (2), 260–264.
14. Hess, P. H., Parker, P. H. Polymers for stabilization of colloidal cobalt particles. *J Appl Polymer Sci* **1966,** *10* (12), 1915–1927.
15. Chantrell, R. W. Agglomerate formation in a magnetic fluid. *J Appl Phys* **1982,** *53* (3), 2742–2744.
16. Vicsek, T., Family, F. Dynamic scaling for aggregation of clusters. *Phys Rev Lett* **1984,** *52* (19), 1669.
17. Popplewell, J., Davies, P., Bradbury, A., Chantrell, R. W. Chain formation in magnetic fluid composites. *IEEE Trans Magn* **1986,** *22* (5), 1128–1130.

18. Miyazima, S., Meakin, P., Family, F. Aggregation of oriented anisotropic particles. *Phys Rev A* **1987**, *36* (3), 1421.
19. Saville, S. L., Woodward, R. C., House, M. J., Tokarev, A., Hammers, J., Qi, B., Shaw, J., Saunders, M., Varsani, R. R., Pierre, T. G. S., et al. The effect of magnetically induced linear aggregates on proton transverse relaxation rates of aqueous suspensions of polymer coated magnetic nanoparticles. *Nanoscale* **2013**, *5* (5), 2152–2163.
20. Saville, S. L., Qi, B., Baker, J., Stone, R., Camley, R. E., Ye, K. L. L. L., Crawford, T. M., Mefford, O. T. The formation of linear aggregates in magnetic hyperthermia. *J Colloid Interface Sci* **2014**, 141–151.
21. Kozissnik, B., Dobson, J. Biomedical applications of mesoscale magnetic particles. *MRS Bull* **2013**, *38* (11), 927–932.
22. Pankhurst, Q. A., Thanh, N. T. K., Jones, S. K., Dobson, J. Progress in applications of magnetic nanoparticles in biomedicine. *J Phys D Appl Phys* **2009**, *42* (22), 224001.
23. Dobson, J. Remote control of cellular behaviour with magnetic nanoparticles. *Nat Nanotechnol* **2008**, *3* (3), 139–143.
24. Molday, R. S., Yen, S., Rembaum, A. Application of magnetic microspheres in labelling and separation of cells. *Nature 268* **1977**, 437–438.
25. Alsberg, E., Feinstein, E., Joy, M. P., Prentiss, M., Ingber, D. E. Magnetically-guided self-assembly of fibrin matrices with ordered nano-scale structure for tissue engineering. *Tissue Eng* **2006**, *12* (11), 3247–3256.
26. Whatley, B. R., Li, X., Zhang, N., Wen, X. Magnetic-directed patterning of cell spheroids. *J Biomed Mater Res* **2013**, 00A: 1–11.
27. Ho, V. H. B., Muller, K. H., Barcza, A., Chen, R., Slater, N. K. H. Biomaterials. *Biomaterials* **2010**, *31* (11), 3095–3102.
28. Souza, G. R., Molina, J. R., Raphael, R. M., Ozawa, M. G., Stark, D. J., Levin, C. S., Bronk, L. F., Ananta, J. S., Mandelin, J., Georgescu, M.-M., et al. Three-dimensional tissue culture based on magnetic cell levitation. *Nat Nanotechnol* **2010**, *5* (4), 291–296.
29. Bratt-Leal, A. M., Kepple, K. L., Carpenedo, R. L., Cooke, M. T., McDevitt, T.C. Magnetic manipulation and spatial patterning of multi-cellular stem cell aggregates. *Integr Biol* **2011**, *3* (12), 1224.
30. Alenghat, F. J., Fabry, B., Tsai, K. Y., Goldmann, W. H., Ingber, D. E. Analysis of cell mechanics in single vinculin-deficient cells using a magnetic tweezer. *Biochem Biophys Res Commun* **2000**, *277* (1), 93–99.
31. Frasca, G., Gazeau, F., Wilhelm, C. Formation of a three-dimensional multicellular assembly using magnetic patterning. *Langmuir* **2009**, *25* (4), 2348–2354.
32. Krebs, M. D., Erb, R. M., Yellen, B. B., Samanta, B., Bajaj, A., Rotello, V. M., Alsberg, E. Formation of ordered cellular structures in suspension via label-free negative magnetophoresis. *Nano Lett* **2009**, *9* (5), 1812–1817.
33. Tseng, P., Judy, J. W., Di Carlo, D. Magnetic nanoparticle–mediated massively parallel mechanical modulation of single-cell behavior. *Nat Methods* **2012**, *9* (11), 1113–1119.
34. Ino, K., Okochi, M., Konishi, N., Nakatochi, M., Imai, R., Shikida, M., Ito, A., Honda, H. Cell culture arrays using magnetic force-based cell patterning for dynamic single cell analysis. *Lab Chip* **2007**, *8* (1), 134.
35. Polte, T. R., Shen, M., Karavitis, J., Montoya, M., Pendse, J., Xia, S., Mazur, E., Ingber, D. E. Nanostructured magnetizable materials that switch cells between life and death. *Biomaterials* **2007**, *28* (17), 2783–2790.

36. Lee, C., Lee, H., Westervelt, R. Microelectromagnets for the control of magnetic nanoparticles. *Appl Phys Lett* **2001**, *79* (20), 3308–3310.
37. Yellen, B., Friedman, G., Feinerman, A. Analysis of interactions for magnetic particles assembling on magnetic templates. *J Appl Phys* **2002**, *91* (10), 8552.
38. Yellen, B. B., Friedman, G. Analysis of repulsive interactions in chains of super-paramagnetic colloidal particles for magnetic template-based self-assembly. *J Appl Phys* **2003**, *93* (10), 8447–8449.
39. Plaks, A., Tsukerman, I., Friedman, G., Yellen, B. Generalized finite-element method for magnetized nanoparticles. *IEEE Trans Magn* **2003**, *39* (3), 1436–1439.
40. Hovorka, O., Yellen, B., Friedman, G. Modeling stability of trapped ferromagnetic nanoparticle chains. *IEEE Trans Magn* **2003**, *39* (5), 2549–2551.
41. Yellen, B., Friedman, G., Feinerman, A. Printing superparamagnetic colloidal particle arrays on patterned magnetic film. *J Appl Phys* **2003**, *93* (10), 7331–7333.
42. Love, J., Urbach, A., Prentiss, M., Whitesides, G. Three-dimensional self-assembly of metallic rods with submicron diameters using magnetic interactions. *J Am Chem Soc* **2003**, *125* (42), 12696–12697.
43. Lee, H., Purdon, A. M., Chu, V., Westervelt, R. M. Controlled assembly of magnetic nanoparticles from magnetotactic bacteria using microelectromagnets arrays. *Nano Lett* **2004**, *4* (5), 995–998.
44. Yellen, B., Friedman, G. Programmable assembly of heterogeneous colloidal particle arrays. *Adv Mater* **2004**, *16* (2), 111.
45. Helseth, L. E., Fischer, T. M., Johansen, T. H. Magnetic structuring and transport of colloids at interfaces. *J Magn Magn Mater* **2004**, *277* (3), 245–250.
46. Sahoo, S., Kontos, T., Schönenberger, C., Sürgers, C. Electrical spin injection in multiwall carbon nanotubes with transparent ferromagnetic contacts. *Appl Phys Lett* **2005**, *86* (11), 112109.
47. Yellen, B. B., Hovorka, O., Friedman, G. Arranging matter by magnetic nanoparticle assemblers. *PNAS* **2005**, *102* (25), 8860.
48. Gunnarsson, K., Roy, P., Felton, S., Pihl, J., Svedlindh, P., Berner, S., Lidbaum, H., Oscarsson, S. Programmable motion and separation of single magnetic particles on patterned magnetic surfaces. *Adv Mater* **2005**, *17* (14), 1730–1734.
49. Furlani, E. Analysis of particle transport in a magnetophoretic microsystem. *J Appl Phys* **2006**, *99* (2), 024912.
50. Furlani, E., Ng, K. Analytical model of magnetic nanoparticle transport and capture in the microvasculature. *Phys Rev E* **2006**, *73* (6), 061919.
51. Hovorka, O., Yellen, B., Dan, N., Friedman, G. Self-consistent model of field gradient driven particle aggregation in magnetic fluids. *J Appl Phys* **2005**, *97* (10), 10Q306.
52. Ge, J., Hu, Y., Yin, Y. Highly tunable superparamagnetic colloidal photonic crystals. *Angew Chem Int Ed* **2007**, *46* (39), 7428–7431.
53. Ge, J., Hu, Y., Biasini, M., Beyermann, W. P., Yin, Y. Superparamagnetic magnetite colloidal nanocrystal clusters. *Angew Chem Int Ed* **2007**, *46* (23), 4342–4345.
54. Park, J.-I., Jun, Y.-W., Choi, J.-S., Cheon, J. Highly crystalline anisotropic super-structures via magnetic field induced nanoparticle assembly. *Chem Commun* **2007**, *47*, 5001–5003.
55. Liu, M., Lagdani, J., Imrane, H., Pettiford, C., Lou, J., Yoon, S., Harris, V. G., Vittoria, C., Sun, N.X. Self-assembled magnetic nanowire arrays. *Appl Phys Lett* **2007**, *90* (10), 103105.

56. Yellen, B. B., Erb, R. M., Halverson, D. S., Hovorka, O., Friedman, G. Arraying nonmagnetic colloids by magnetic nanoparticle assemblers. *IEEE Trans Magn* **2006**, *42* (10), 3548–3553.

57. Erb, R. M., Yellen, B. B. Concentration gradients in mixed magnetic and non-magnetic colloidal suspensions. *J Appl Phys* **2008**, *103* (7), 07A312.

58. Erb, R. M., Sebba, D. S., Lazarides, A. A., Yellen, B.B. Magnetic field induced concentration gradients in magnetic nanoparticle suspensions: Theory and experiment. *J Appl Phys* **2008**, *103* (6), 063916.

59. Erb, R. M., Son, H. S., Samanta, B., Rotello, V. M., Yellen, B. B. Magnetic assembly of colloidal superstructures with multipole symmetry. *Nature* **2009**, *457* (7232), 999–1002.

60. Li, K. H., Yellen, B. B. Magnetically tunable self-assembly of colloidal rings. *Appl Phys Lett* **2010**, *97* (8), 083105.

61. Khalil, K. S., Sagastegui, A., Li, Y., Tahir, M. A., Socolar, J. E. S., Wiley, B. J., Yellen, B. B. Binary colloidal structures assembled through ising interactions. *Nat Commun* **2012**, *3*, 794–798.

62. Ozdemir, T., Sandal, D., Culha, M., Sanyal, A., Atay, N. Z., Bucak, S. Assembly of magnetic nanoparticles into higher structures on patterned magnetic beads under the influence of magnetic field. *Nanotechnology* **2010**, *21* (12), 125603.

63. Ruan, G., Vieira, G., Henighan, T., Chen, A., Thakur, D., Sooryakumar, R., Wintert, J. O. Simultaneous magnetic manipulation and fluorescent tracking of multiple individual hybrid nanostructures. *Nano Lett* **2010**, *10* (6), 2220–2224.

64. Fragouli, D., Buonsanti, R., Bertoni, G., Sangregorio, C., Innocenti, C., Falqui, A., Gatteschi, D., Cozzoli, P. D., Athanassiou, A., Cingolani, R. Dynamical formation of spatially localized arrays of aligned nanowires in plastic films with magnetic anisotropy. *ACS Nano* **2010**, *4* (4), 1873–1878.

65. Chen, A., Vieira, G., Henighan, T., Howdyshell, M., North, J. A., Hauser, A. J., Yang, F. Y., Poirier, M. G., Jayaprakash, C., Sooryakumar, R. Regulating brownian fluctuations with tunable microscopic magnetic traps. *Phys Rev Lett* **2011**, *107* (8), 087206.

66. Rapoport, E., Beach, G. S. D. Dynamics of superparamagnetic microbead transport along magnetic nanotracks by magnetic domain walls. *Appl Phys Lett* **2012**, *100* (8), 082401.

67. Lim, J., Tan, D. X., Lanni, F., Tilton, R. D., Majetich, S. A. Optical imaging and magnetophoresis of nanorods. *J Magn Magn Mater* **2009**, *321* (10), 1557–1562.

68. Demirörs, A. F., Pillai, P. P., Kowalczyk, B., Grzybowski, B. A. Colloidal assembly directed by virtual magnetic moulds. *Nature* **2013**, *503* (7474), 99–103.

69. Erb, R. M., Libanori, R., Rothfuchs, N., Studart, A. R. Composites reinforced in three dimensions by using low magnetic fields. *Science* **2012**, *335* (6065), 199–204.

70. Urbach, A. R., Love, J. C., Prentiss, M. G., Whitesides, G. M. Sub-100 Nm confinement of magnetic nanoparticles using localized magnetic field gradients. *J Am Chem Soc* **2003**, *125* (42), 12704–12705.

71. Korth, B. D., Keng, P., Shim, I., Bowles, S. E., Tang, C., Kowalewski, T., Nebesny, K. W., Pyun, J. Polymer-coated ferromagnetic colloids from well-defined macromolecular surfactants and assembly into nanoparticle chains. *J Am Chem Soc* **2006**, *128* (20), 6562–6563.

72. Xue, X., Furlani, E.P. Template-assisted nano-patterning of magnetic core–shell particles in gradient fields. *Phys Chem Chem Phys* **2014**, *16* (26), 13306.

73. Henderson, J. R., Crawford, T. M. Repeatability of magnetic-field driven self-assembly of magnetic nanoparticles. *J Appl Phys* **2011**, *109* (7), 07D329.
74. Henderson, J., Shi, S., Cakmaktepe, S., Crawford, T. M. Pattern transfer nanomanufacturing using magnetic recording for programmed nanoparticle assembly. *Nanotechnology* **2012**, *23* (185304), 185304.
75. Ye, L., Terry, B., Mefford, O. T., Rinaldi, C. All-nanoparticle concave diffraction grating fabricated by self-assembly onto magnetically-recorded templates. *Opt Express*, Volume 21 **2013**, 1066–1075.
76. Wang, S. X., Taratorin, A. M. *Magnetic Information Storage Technology*. San Diego, CA: Academic Press, 1999.
77. Ye, L., Pearson, T., Dolbashian, C., Pstrak, P. Magnetic-field-directed self-assembly of programmable mesoscale shapes. *Adv Funct*, Volume 26 **2016**, DO 10.1002/adfm.201504749.
78. Coey, J. M. D. Noncollinear spin arrangement in ultrafine ferrimagnetic crystallites. *Phys Rev Lett* **1971**, *27* (17), 1140–1142.
79. Krycka, K. L., Booth, R. A., Hogg, C. R., Ijiri, Y., Borchers, J. A., Chen, W. C., Watson, S. M., Laver, M., Gentile, T. R., Dedon, L. R., et al. Core-shell magnetic morphology of structurally uniform magnetite nanoparticles. *Phys Rev Lett* **2010**, *104* (20), 207203.
80. Maiti, S., Chen, H.-Y., Chen, T.-Y., Hsia, C.-H., Son, D. H. Effect of surfactant and solvent on spin-lattice relaxation dynamics of magnetic nanocrystals. *J Phys Chem B* **2013**, *117* (16), 4399–4405.
81. del Castillo, V.-D., Rinaldi, C. Effect of sample concentration on the determination of the anisotropy constant of magnetic nanoparticles. *IEEE Trans Magn* **2010**, *46* (3), 852–859.

9

Applications of Magnetic Nanoparticles in Tissue Engineering and Regenerative Medicine

James R. Henstock, Hareklea Markides, Hu Bin,
Alicia J. El Haj, and Jon Dobson

CONTENTS

9.1 Introduction 205
9.2 Tissue Engineering and Regenerative Medicine 206
9.3 Mechanotransduction 208
9.4 Targeting Mechanoreceptors on the Cell Membrane 208
9.5 Magnetic Energy Transfer 211
9.6 Applications of Magnetic Nanoparticles in TERM 211
9.7 Magnetic Activation of Receptor Signaling/Magnetic Ion
    Channel Activation Technology 216
9.8 Conclusion 222
References 223

9.1 Introduction

Biomolecule-functionalized magnetic nanoparticles (MNPs) can be targeted to specific cell-surface receptors and then those receptors remotely activated by transducing forces or energy from a remote, external magnetic field (Dobson and St. Pierre, 1996; Cartmell et al., 2002, 2005; Dobson, 2008; Dobson et al., 2006; Hughes et al., 2008; Henstock et al., 2014; Creixell et al., 2011; Mannix et al., 2008; Henstock and El Haj, 2015; El Haj et al., 2015). This remote control of receptor activity has numerous applications, such as the direct mechanical activation of stretch-activated ion channels (Hughes et al., 2005a, 2006, 2007, 2008), extracellular matrix binding proteins (Cartmell et al., 2002; Hughes et al., 2007; Henstock et al., 2014), surface kinases (Hu et al., 2014), and G-protein-coupled receptors (Rotherham and El Haj, 2015), while the ability to control the movement of a weakly interacting ligand allows for a switchable regulation of the interface, and thus receptor activation.

Because of the way the nanoparticles convey mechanical forces directly to their designated target, this approach has clear applications in delivering mechanical stimuli directly to the surface mechanoreceptors of cells—and without having to deform the extracellular scaffold the nanoparticles provide a mechanism by which cells on soft or fragile biomaterials can be mechanically loaded (Cartmell et al., 2005; Dobson et al., 2006; Wimpenny et al., 2012; Henstock et al., 2014). Many cells are shown to respond to mechanical stimuli by producing stronger, more effective extracellular matrix, and perhaps the most clinically relevant of these are mesenchymal stem cells (MSCs), which are being evaluated in a host of cell-based strategies for orthopedic tissue engineering. Substantial evidence suggests that mechanical stimuli play a role in the osteoblastic differentiation of MSCs, interacting with and amplifying signaling cascades from growth factors such as bone morphogenetic proteins (BMPs) (Henstock et al., 2014; Kopf et al., 2012; Schwarz et al., 2013). As a result, the presence or absence of mechanical stimulation in the biochemically complex environment during stem-cell differentiation and tissue synthesis can have a large impact on the quality and quantity of bone formed, potentially affecting the clinical outcome of both conventional and tissue-engineered treatments for nonunion and the integration of articular prostheses.

9.2 Tissue Engineering and Regenerative Medicine

Tissue engineering and regenerative medicine (TERM) describe a variety of techniques for restoring function to lost, damaged, diseased, or aged tissues. Broadly, the approaches are similar, but while tissue engineering focusses on the generation of replacement tissues using cell-seeded biomaterial scaffolds, regenerative medicine describes a host of other techniques, often using stem-cell therapies to restore tissue functionality to a prepathological state, and often *in vivo*.

In all of these therapies, there is a significant focus toward a directed outcome, that is, that the scaffold can degrade at a predetermined rate and crucially that the cell component of the therapy can be controlled, usually by means of interaction with the biomaterial, or a spatiotemporal control over the release of growth factors. The cell component meanwhile, is expected to differentiate, proliferate, and produce an effective extracellular matrix—effectively recapitulating a normal tissue repair process—albeit with technological assistance.

However, a key element to normal repair or tissues is crucially missing in most of these strategies—mechanical stimulation (Vogel and Sheetz, 2009). Cells exist in a dynamic and complex mechanical environment in which cues from the normal motion of the body help to orchestrate tissue homeostasis, allowing cells to integrate signals from growth factors and cytokines to optimize their production of extracellular matrix. In a complex wound or tissue defect containing a biomaterial scaffold, these mechanical cues are likely to be abnormal and compounded by the fact that biomaterials are either mechanically too stiff (ceramics, metals, some plastics), leading to stress shielding, or too soft (hydrogels) to match the normal tissue. Additionally, patients with a requirement for TERM therapies are likely to have impaired mobility, because of age, complex trauma, or underlying pathologies, undermining the potential for rehabilitation physiotherapy and leaving the therapeutic cells in a mechanically stagnant environment.

There is therefore a significant requirement to solve this challenge and find translatable technologies that can provide appropriate mechanical stimuli to cells and tissues undergoing regeneration. Using targeted, functionalized MNPs, it is possible to deliver stimulatory mechanical forces directly to cells.

Mechanical cues play important roles during embryogenesis and the continuous process of remodeling in mature tissues (Mammoto and Ingber, 2010; Nowlan et al., 2010; Khan and Scott, 2009). Mechanical stimulation has been shown to have numerous effects on the molecular machinery of cells such as opening certain ion channels, propagating phosphorylation cascades in mechanotransduction signaling pathways, exposing binding sites in proteins, and activating bound or latent growth factors (Nilius and Honore, 2012; Mobasheri et al., 2012; Storch et al., 2012; Albro et al., 2012; Miyanozo et al., 2008). As such, precise control of the delivery of mechanical cues to specific mechanosensitive cellular receptors holds great potential for targeted effects, similar in principle to pharmacological approaches, but via a rapid, controllable physical mechanism.

Current methods in TERM for the application of mechanical stimulation to cells and tissues *in vitro* culture use bioreactors and have some significant drawbacks, particularly in the lead time that the cell-seeded scaffolds have to be cultured for within the engineered tissue, which is usually several weeks (Hung et al., 2004; Martin et al., 2007). Magnetic ion channel activation (MICA) and magnetic activation of receptor signaling (MARS) technologies employ MNPs combined with magnetic fields to deliver highly localized mechanical/energetic cues to specific cell-surface receptors. Magnetomechanical stimulation and activation of cell receptors has been successfully employed to activate ion channels (both targeted and general mechanoactive channels) as well as specific cell-surface receptors, enabling remote control of specific cell signaling pathways (Hughes et al., 2008; Dobson, 2008; Huang et al., 2010; Mannix et al., 2008). The MICA and MARS technologies allow specific cell receptors to be remotely activated and their associated signaling cascades controlled, presenting both researchers and clinicians with enticing possibilities for remote control of cell function.

9.3 Mechanotransduction

Mechanotransduction is the conversion of mechanical stimuli into biochemical signals, allowing cells to sense their surrounding physical environment and react to it appropriately (Storch et al., 2012). Mechanotransduction plays an important role in cell proliferation, differentiation, migration, and apoptosis—processes that are all necessary for tissue development, homeostasis, and healing (for further details, see review by Mammoto and Ingber, 2010). During both development and remodeling, cells are subjected to various forms of mechanical stimulation, which shapes and regulates a large array of physiological processes. For example, bone formation and remodeling are regulated by fluid flow and compressive loading; the vascular system is influenced by pressure and shear stresses caused by the pumping of blood; the sensory neural system receives pressure inputs that are converted into biochemical and then electrical signals for hearing and touch (Papachroni et al., 2009).

Aberrant mechanotransduction processes are involved in many pathological conditions including oncogenesis (Janmey and Miller, 2011). The mechanisms underlying the conversion of mechanical cues into biochemical signals have therefore attracted great attention but there is a substantial number of unresolved questions regarding the types and ranges of forces to which specific cells are sensitive, and the expression of the transduction machinery, signaling pathways, and their interactions that facilitate a coherent response.

Mechanosensation can be described at a series of hierarchical levels, from specific sensitive organs (such as the ear) through to specialized cell types in other tissues such as the osteocyte. A key role of the osteocyte is the sensation of bone loading; and specialized structures called dendritic processes within the canalicular space both link cells and amplify mechanical signals due to the pressurized flow of fluid along the process. An array of sensitive proteins and complexes are present on the surface of the cell, while the extracellular matrix and intracellular cytoskeleton connect through integrins at the cell membrane. At the scale of individual proteins and biological macromolecules, picoNewton (pN)-scale mechanical forces have been shown to have effects on the chemical bonds that comprise the mechanosensitive proteins (Paluch et al., 2015; Jurchenko and Salaita, 2015).

9.4 Targeting Mechanoreceptors on the Cell Membrane

A variety of cell-surface receptors are potential targets for magnetic activation to initiate downstream signaling. The forces applied through external fields can mechanoactivate receptors through multiple routes. Tensile forces applied on ligand-receptor complexes can shorten the lifetime of the covalent bonds by accelerating their dissociation rate (Evans and Calderwood, 2007). Mechanoresponsive proteins undergo functional conformational transitions in response to tension that alter binding properties or enzymatic functions (Sawada et al., 2006; Shemesh et al., 2005). The intracellular protein Src is one such type of protein whose mechanical activation depends on mechanical forces on integrins (Na et al., 2008).

Mechanical forces can also change the sensitivity of disulfide bonds in response to the redox state of the cell microenvironment. This could be explained by the chemomechanical kinetic model, in that force potentially lowers the activation barrier proportionally to the difference in a single internuclear distance between the ground and transition states projected on the force vector (Yang et al., 2009).

Nanoparticles also can be specifically targeted to individual receptors on cells by functionalizing the nanoparticle surface with an antibody, ligand, or binding motif. Chen et al. (2001) have reported that twisting integrin receptors with arginine-glycine-aspartic acid (RGD)-containing peptide-coated beads via magnetic twisting cytometry enhanced endothelin-1 gene expression by more than 100%. The underlying mechanism was postulated to be the linkage of integrin-cytoskeleton involving stretch-activated ion channel and myosin adenosine triphosphatase (ATPase). Fass et al. utilized anti-β1 integrin antibody-coated magnetic beads to induce neurite elongation (Fass and Odde, 2003). They showed that neurites subjected to constant force elongated at variable instantaneous rate and switched abruptly between elongation and retraction, which was similar to the process that happened *in vitro*. The low force threshold for elongation was reported to be 15–100 pN. This application may be applied to neural tissue engineering.

Epidermal growth factor (EGF)-coated magnetic beads have been used for local delivery of EGF and the downstream signaling cascade of EGF receptor activation including actin polymerization and extracellular-signal-regulated kinase (ERK) activation (Kempiak and Segall, 2004). The authors found whether a region would be affected or not was dependent on the receptor's distance from the region and the receptor's activation level due to activation binding. In the model employed by the authors, amoeboid chemotaxis would be kept within propagation distances of ~1 µm from the receptor for actin polymerization and at least 10 µm for activated ERK. Similar technique can potentially be applied to local stimulation of growth factor receptors in certain clinical situations. In 2008, multiligand functionalized nanoparticles were first demonstrated to be a versatile tool for receptor dimerization or clustering and the subsequent downstream signaling cascade activation (Mannix et al., 2008).

Another key target for this nanoparticle binding approach is the targeting of ion channels within the cell membrane. An example is the mechanosensitive ion channel TWIK-Related K+ Channel 1 (TREK1), a stretch-activated potassium channel that is gated in normal cell physiology by an intracellular

C-terminus tail region, which can associate with the intracellular side of the cell membrane. Hughes, Dobson, and El Haj utilized magnetic particles for applying highly localized forces to distinct regions of the TREK1 ion-channel structure, and successfully changed whole-cell currents when particles were targeted toward the extended extracellular loop region of TREK1 (Hughes et al., 2008). Responses were absent when particles were coated with RGD peptide other than TREK1 antibody or when magnetic fields were applied in the absence of magnetic particles (Figure 9.1).

Wnt signal transduction pathways have also been targeted which recognize the canonical and noncanonical pathways, and are known to play crucial roles in mediation of signaling (Logan and Nusse, 2004; Kurayoshi et al., 2007). The

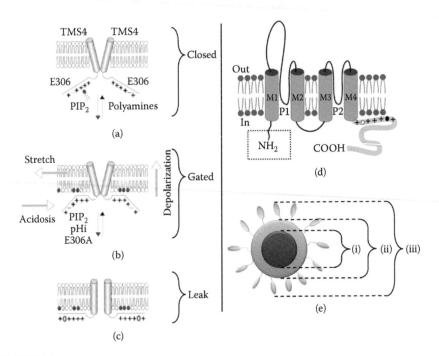

FIGURE 9.1
Gating of the mechanosensitive (membrane stretch-activated) ion channel TWIK-Related K+ Channel 1 (TREK1) is regulated in normal cell physiology by an intracellular C-terminus tail region, which can associate with the intracellular side of the cell membrane, resulting in three conformational activation states, closed (a), gated (b), and leaky (c). Deformation of the membrane caused by mechanical load therefore results in ion channel activation in either of the latter conformational states. By functionalizing the surface coating of MNPs with an antibody against the N-terminus (d, indicated) the activity of the ion channel can be remotely controlled by a variable, remote magnetic field, which provides the requisite force to activate the channel. The nanoparticles (e) consist of an superparamegnetic iron oxide (SPIO) core (i) surrounded by a biocompatible dextran coating (ii), which can be functionalized with binding motifs, in this case an antibody (iii) against the N-terminal domain of TREK1. Images (a through c). (Adapted from Chemin, J. et al., *EMBO J*, 24, 44–53, 2005; Sandoz, G. et al., *Proc Natl Acad Sci U S A*, 108, 2605–2610, 2011.)

nanoparticle-based approach has been used to offer a potential clinically translatable mechanism to regulate the pathway at either the receptor level (Rotherham and El Haj, 2015) or by constructing a standardized synthetic Wnt-like ligand which can bind into the receptors using a common motif, but which only activates Wnt signaling when force is applied using the magnet (Laeremans et al., 2011).

In addition to these examples, there are a number of other molecular pathways in which having direct control of ligand binding or receptor activation may have significant clinical or research applications (e.g., Hu et al., 2013, 2014).

9.5 Magnetic Energy Transfer

In addition to applying translational forces to particles via magnetic field gradients or twisting forces on magnetically blocked particles (see Pankhurst et al., 2009), recent work has demonstrated the possibility of transferring energy from a radiofrequency (RF) magnetic field to MNPs attached to surface receptors. The principles are similar to those used in magnetic fluid hyperthermia. In 2011, Creixell et al. were able to target EGF receptors and kill cancer cells using RF fields without any measured rise in temperature—a process that may be due to activation of an apoptotic EGFR pathway (Creixell et al., 2011). In 2010, Huang and Pralle demonstrated *in vivo* RF activation of MNPs targeting neuronal receptors, which influenced the behavior of *Caenorhabditis elegans* (Huang et al., 2010). Clearly, this is a technique with potential future application as RF fields have the potential to transfer energy to particle/receptor targets deep within the body, inducing conformational changes and activating cell signaling.

9.6 Applications of Magnetic Nanoparticles in TERM

Several studies have demonstrated the multiple applications of MNPs in tissue engineering including cell labeling and tracking, as well as cell and biochemical molecular patterning, and tissue fabrication (Markides et al., 2013; Harrison et al., 2016; Perea et al., 2007; Kilgus et al., 2012; Ito et al., 2007). Tracking, visualizing and controlling the behavior of lab-grown stem cells in the body are key challenges in regenerative medicine. It is possible to achieve all three of these goals by labeling the cells with MNPs and using technologies which exploit the unique properties of the nanoparticles.

Many of the signaling processes involved in stem-cell differentiation and tissue healing are finely tuned, with complex temporal and spatial regulation of signaling ligands, and interconnected pathways that serve to integrate different inputs into a coherent response (Lin and Hankenson, 2011), and so broad pharmacological approaches to stimulating tissue regeneration may not always be appropriate. In practice, an adaptable control method is required which can be used to remotely control these signal-transducing receptors from a distance (i.e., by clinicians from the bedside) and which can activate therapeutic pathways when required—for example in response to patient healing rates or to modulate the local effects of a systemic drug treatment.

Among other authors, we have demonstrated that by functionalizing the surface of a magnetic nanoparticle with specific ligands, we can target the nanoparticles to specific proteins or receptors on the cell surface, which then provides direct control of the receptor—in essence an electromagnetic "switch" (Figure 9.1). We have demonstrated that the approach is particularly appropriate for activating mechanoreceptors, as these receptors are fine-tuned to respond to the direct mechanical loading provided by the magnetically activated nanoparticle. Indeed, our group has also previously reported on mechanoactivation of integrins in this same way, by functionalizing the nanoparticle surface with the conserved RGD tripeptide motif present in the integrin-binding domains of extracellular matrix proteins, thereby proving stimuli to cells as though the surrounding matrix were being deformed by external force and stimulating the downstream processes of mechanotransduction (Cartmell et al., 2002; Hughes et al., 2007; Henstock et al., 2014).

Other recent studies have demonstrated that iron oxide MNPs could be utilized for magnet-guided cell therapy to augment cell homing and retention at the site of tissue repair, particularly using magnetic resonance imaging (MRI) magnets to guide particles and MNP-loaded cells through deep tissue (Riegler et al., 2013; Muthana et al., 2015). Circulating endothelial progenitor cells (EPCs) are involved in vascular re-endothelialization and postischemic neovascularization, and subsequently the successful delivery of EPCs will be important for cell therapies. A variety of studies have shown that external magnetic fields can guide transplanted cells to the damaged tissues and enhance therapeutic effects (summarized in Table 9.1), with significant enhancement of up to 6-fold over unlabeled, untrapped controls. Kwan et al. explored the effect of magnetic field strength and time on the degree of cell migration in response to an external magnetic field. Applying several *in vitro* transwell migration systems, the group demonstrated that MNP-loaded cells were able to migrate toward the magnetic field in a dose-dependent manner both under static and dynamic conditions. This was further tested in an *in vivo* rabbit model of tendinopathy where MNP-loaded cells were shown to have successfully migrated to the magnetized scaffold implanted within the Achilles tendon sheath while in rabbits treated with unlabeled cells they appeared only on the surface of the tendon and did not migrate to

TABLE 9.1

Several Studies Have Shown That Magnetic Trapping of Cells Is a Viable Route Toward Optimizing Delivery in Some Therapies, Particularly for Vascular Applications in Which a Significant Shear Force Must Be Overcome to Localize Circulating Cells into a Target or Wound Site

Tissue/Model	Cell Type	Summary	Author
Cerebral (ischemia)	EPCs	EPCs homing was significantly increased in the ischemic hemisphere of the brain in mice.	Li et al. (2013)
Vascular (model)	MSC	6-fold increase in cell retention in an *in vitro* flow system.	Kyrtatos et al. (2009)
Vascular (injury)	MSC	5-fold increase in cell retention at site of vascular injury.	
Vascular	MSC	6-fold increase in MSC retention; reduction in restenosis following balloon angioplasty in a rabbit model.	Riegler et al. (2013)
Vascular	MSC	Tunable magnetic pole for deep magnetic capture of cells for spatial targeting therapeutics.	Huang et al. (2010)
Vascular	Endothelial cells	Enhanced efficiency of endothelial cells for treating vascular diseases in a mouse model.	Hofmann et al. (2009)
Vascular (model)	MSC	Volume of trapped cells increased proportionally with MNP concentration and decreased with increasing flow rate.	El Haj et al. (2015)
Subcutaneous wound	MSC	An implanted magnet induced the trapping of MNP-loaded MSCs in a subcutaneous wound model in nude mice.	El Haj et al. (2015)
Liver	Huh7	Enhanced perivascular homing in the liver (mouse).	Luciani et al. (2009)
In vitro positioning	MSCs/ cardiomyocytes	Soft iron tips (~500 μm diameter) were utilized to generate magnetic fields for local gene targeting and positioning of MSCs or cardiomyocytes.	Kilgus et al. (2012)
Tendon rabbit model	Adipose-derived MSCs	MNP-loaded cells demonstrated improved migration toward the magnetized PHBXX tendon scaffold *in vivo*.	Thomas D. Kwan (2015)

Note: Conversely, other applications have focused on uses of the technology for more precise targeting, for example, within the brain or to create patterned networks of cells.

the damaged site (Kwan and El Haj, unpublished) (Figure 9.2). Equally, the potential for using the magnetic trapping technology for precise spatial pattern of the labeled cells was explored (Kilgus et al., 2012).

Another approach in TERM is for applications in engineering three-dimensional (3D) tissues which utilize the combination of magnetic particles

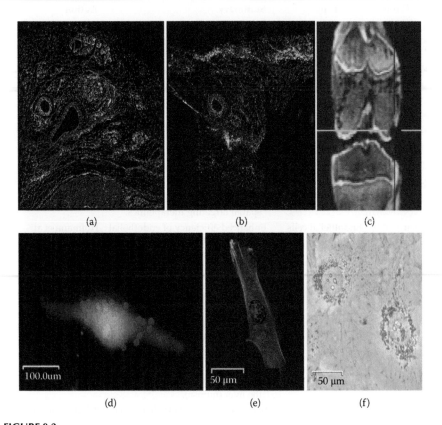

(a)   (b)   (c)

(d)   (e)   (f)

FIGURE 9.2
Labeling therapeutic cell populations with MNPs allows for cells to either be guided with an external magnetic field toward the sight of injury or encouraged to migrate toward a magnetic field, for example a magnetized scaffold implanted at the sight of injury. In this example (a and b), a magnetized scaffold for tendon repair was shown to improve the migration of implanted MNP-loaded cells (fluorescently membrane labeled in red) toward the repair site in a rabbit. The cells were able to penetrate the tendon sheath (a) while unlabeled cells are only present upon the surface of the tendon sheath (b). Nanoparticles have commonly been used as MRI contrast agents, and in this example (c), prelabeling therapeutic cells with MNPs allows the cells to be imaged in the defect site, in this case the lateral condyle of a rat femur. The coating of nanoparticles can also be made fluorescent in addition to being functionalized with targeting ligands, allowing the location of the nanoparticles on the cell to be determined, for example when binding integrins on the cell surface with an RGD-motif (d). These nanoparticles can be easily distinguished in histology and cell culture by staining their iron cores with Prussian blue (e). (Images (a) and (b) from Kwan and El Haj, unpublished; (c) from Markides, H. et al., *Stem Cell Res Ther*, 4, 126, 2013; (d) from Hughes, S. et al., *J Biomech*, 40, S96–S104, 2007; (e) from Pašukonienė, V. et al., *Medicina*, 50, 237–244, 2014).

and magnetic fields to direct the assembly of 3D cell clusters or aggregates. M-gels formed by MNPs and cell-encapsulation hydrogels can be utilized as building blocks to form 3D tissue constructs by spatially controlling the magnetic field (Xu et al., 2011, 2012). In another study, magnetic nanowires were used to control the spatial positioning of NIH-3T3 mouse fibroblasts cells in combination with a micromagnet array (Tanase et al., 2005). In this system, cell-pattern efficiency and speed were significantly enhanced when using fluid flow to bring the cells onto the arrays.

Highly efficient fibroblast and MSC-seeding have been achieved through magnetite cationic liposome and magnetic forces (Shimizu et al., 2007a, b). A magnetic force-directed biodynamic substrate was employed for the manipulation of extracellular matrix (ECM) deposition with cell-adhesive fragments of fibronectin-functionalized 100-nm prefabricated albumin nanoparticles (Pereira et al., 2007). RGD-functionalized superparamagnetic microbeads can act as a tool for cell patterning (Ito et al., 2007) and a control over apoptosis could be established when a magnetic field was applied or withdrawn (Polte et al., 2007). This method could be used as a reconfigurable interface, allowing for cell selection and replacement to form a purified cell group or complex structure.

Vascular network formation was also reported to be facilitated by magnetic activation. Wilhelm et al. moved magnetically-labeled EPCs and modified their organization in Matrigel™ both *in vivo* and *in vitro* (Wilhelm et al., 2007). After magnetically labeling porcine endothelial cells with superparamagnetic iron oxide (SPIO) microspheres, a flexible magnetic sheet was used to rapidly and efficiently cover and retain cells in the surface of a vascular graft in the presence of blood flow (Pislaru et al., 2006). Similar magnetic manipulations have reportedly been applied in vascular tissue engineering (Perea et al., 2006, 2007). Antibody-conjugated magnetoliposomes combined with magnetic force guidance proved to be a useful tool for keratinocytes, cardiomyocytes, MSC sheets and liver-like 3D tissue construction (Ito et al., 2005, 2008).

Magnetic force-induced cell levitation is an alternative approach to cell patterning. In 2010, the first report was published to describe a 3D tissue culture based on magnetic levitation of cells (Souza et al., 2010; Chapter 10, this volume). Cells were loaded with magnetic iron oxide nanoparticles and then exposed to an external magnetic field to form clusters. The geometry of the resulting cell mass could be manipulated and different types of cells be organized to form cocultured clusters by spatially controlling the magnetic field. By using this magnetic levitation method to culture human glioblastoma cells, the authors found that the cells expressed a similar protein profile to those obtained from human tumor xenografts. A similar technique has been applied to the culture of KB tumor cells, adipose tissue, and heterogenous cultures of bronchiole cells in 3D clusters (Lee et al., 2006, 2011; Daquinag et al., 2013; Tseng et al., 2013). In the bronchiole coculture model, four types of human cells (endothelial, smooth muscle, fibroblast, and epithelial cells) were rapidly assembled by using magnetic levitation, which continued to exhibit a similar phenotype and extracellular matrix native to the tissue (Tseng et al., 2013).

Being both magnetic and relatively large (by molecular standards), nanoparticles also offer the opportunity to visualize cells and tissues. Using advanced, modern techniques it is becoming possible to translate the visualization methodologies from benchtop imaging to fully scaled *in vivo* models, and ultimately to clinical applications, perhaps providing a resolution to a long-standing question in stem-cell translational medicine: Where do cells go once introduced to the body? (Shapiro et al., 2015). By adapting existing high-resolution MRI and whole-body imaging, MNPs may ultimately offer a coherent method for both controlling the behavior and the location of stem cells in the body, providing a precisely (remote) controlled cell therapy (Kobayashi et al., 2008).

9.7 Magnetic Activation of Receptor Signaling/ Magnetic Ion Channel Activation Technology

The theoretical principles and prototype design of MARS/MICA was proposed by Dobson and El Haj in 2002 and were based on theoretical work by Dobson and St. Pierre in 1996 (Dobson and St. Pierre, 1996; Cartmell et al., 2002). A magnetic force bioreactor (MFB) was designed to apply mechanical stimulation directly to mechanosensitive cell membrane molecules using forces generated on MNPs. The MFB consists of a vertically oscillating, high-gradient magnet array and a control unit. The magnet array can fit into a standard bioincubator cabinet for *in vitro* studies. The frequency and amplitude of the oscillations of the array are controlled via a computerized stepper motor system. Tailoring the functionalization of MNPs and magnetic array movement, varying targets and amplitude of forces can be achieved. Initially, it was used to stimulate mechanosensitive ion channels and integrins. Later on, the application was extended to other mechanically sensitive molecules on cell membrane, such as growth factor receptors (Figure 9.3).

The effects of magnetic fields on biological systems and human health have been studied for many years. In 1992, Kirschvink developed a model linking biogenic magnetite particles implanted into the human brain with mechanosensitive membrane ion gates in extremely low frequency (ELF) alternating magnetic fields (Kirschvink et al., 1992). Later, Dobson and St. Pierre showed that pulsed fields, square waves, and DC magnetic fields were also capable of opening the membrane gates and influencing relative neurophysiological processes (Dobson and St. Pierre, 1996). Moderate-strength magnetic fields (1–1000 mT) were shown to be able to influence the activity of ion channels embedded in phospholipid membranes. For example, such fields (~80 mT)

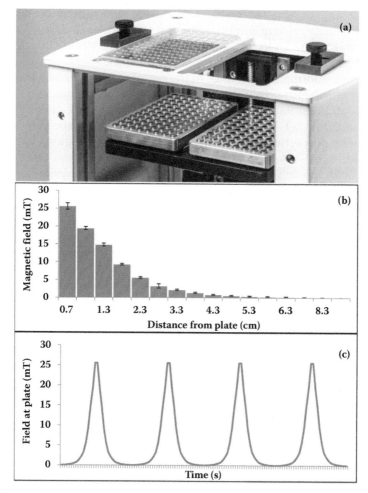

FIGURE 9.3
For cell culture applications, the magnetic ion channel activation (MICA) bioreactor can be used to deliver a magnetic field of variable strength to cells grown in standard culture plates (a). The array of permanent magnets underneath the culture plate is moved vertically by a stepper motor such that proximity of the magnets to the cells on top of the bioreactor will magnetize the SPIO nanoparticles. Moving the array downwards (b) results in a rapid attenuation of the field, allowing precise control of the force applied to the nanoparticle, and therefore transduced to the cell-surface target, for example, the TREK1 ion channel. In usual operation mode, the motor moves the array in a reciprocating manner, delivering an intermittent activation force on the nanoparticle-complex at an experimentally defined frequency, for example, 1 Hz to simulate physiological mechanical loading or routine exercise (c).

were found to influence the open probability and single-channel kinetics of mechanosensitive channel large conductance (MscL) exposed to negative pipette pressure (Hughes et al., 2005a). Many reviews have deliberated these applications (Pankhurst et al., 2003, 2009; Colombo et al., 2012).

Time-varying external magnetic fields used in magnetic actuation for TERM are usually set to mimic physiological frequencies (1–3 Hz). A translational force is generated locally due to the attraction of MNPs along the magnetic field gradient, which stretches the cell membrane or the cytoskeleton, or induces a conformational change in the receptor. Therefore, by tuning the properties of MNPs, surface functionalization and magnetic field strength and gradient, the suitable mechanical stress can be applied directly to cells cultured in 2D or 3D. Moreover, incubating cells with particles with various magnetic properties and seeding them in different regions within the 3D scaffold, a spatial variation of force can be perceived based on the same magnetic field. As such, the differentiated mechanical stimulation of 3D constructs can be realized. Furthermore, MFB are capable of easy scale-up to meet the need for high-throughput culture of multiple constructs following the development of tissue engineering.

Based on a lab-designed MFB, osteoblasts were found to respond instantly to stimulation via different magnetic particles under the same magnetic field and adapted to the applied force following extended periods of magnetic activation culture and exhibited enhanced osteoblasts markers (Hughes et al., 2007).

Glossop and Cartmell examined the effect of human mesenchymal stem cells (hMSCs) exposed to forces generated through magnetic particles combined with an oscillating magnetic field (maximum field strength 90 mT) and uniaxial tensile strain (3% cell elongation) (Glossop and Cartmell, 2010). In contrast to one earlier study, which showed that a mitogen-activated protein kinase (MAP3K8) and interleukin-1β (IL-1B) in MAPK signaling played an important role in the response of hMSCs to fluid shear stress, the follow-up study failed to detect any MAP3K8-related signaling activation, represented by the gene expression change of MAP3K8 or IL-1B, over a 24-hour period subsequent to1-hour force exposure. It is proposed that the activation of specific mechanosensitive pathways allows MSCs to discriminate between different types of mechanical stimuli and react appropriately (Table 9.2).

Membrane hyperpolarization of hMSCs after exposure to dynamic force (6 pN per particle) targeting-integrins was reported by (Kirkham et al., 2010). This local force was generated by RGD-coated ferromagnetic microparticles combined with an external magnetic field. It is suggested that big potassium (BK) channels and intracellular calcium release-mediated cell membrane hyperpolarization responses upon the application of magnetic force.

Kanczler et al. investigated the potential of remote magnetic field activation application to MNP-tagged TREK1 and RGD on the cell membrane of hMSCs to direct their differentiation toward an osteochondral lineage *in vitro* and *in vivo*. hMSCs were cultured in monolayer over a 7-day period (Kanczler et al., 2010). The hMSCs encapsulated into alginate/chitosan microcapsules were cultured both *in vitro* and *in vivo* with daily exposure to a magnetic field over a 24-day period. Several osteochondral gene markers were increased in groups exposed to the forces, as was collagen deposition *in vivo*. These

TABLE 9.2

Activation of Specific Targets on the Cell Surface, Membrane, or Immediate Extracellular Matrix Has Been Successfully Reported by Several Research Groups, Including the Authors

Target	Cell	Effects	Author
Friz (Wnt receptor)	Human MSC	Activation of the Wnt/β-catenin signaling pathway.	Rotherham and El Haj (2015)
Integrins (RGD⁻)	Human MSC	Activated intracellular Ca signaling.	Hughes et al. (2007)
Integrins (RGD⁻)	Human MSC in mouse dorsum	Upregulation of COL1 and COL2.	Kanczler et al. (2010)
TREK1 ion channel	Human MSC	Changes in whole-cell currents consistent with TREK1 activity.	Hughes et al. (2008)
TREK1 ion channel	Human MSC microinjected into chick femur	Induced endochondral ossification at injection/activation site. Upregulation of osteogenic gene expression.	Henstock et al. (2014)
TREK1 ion channel	Human MSC in collagen hydrogel	Differentiation and substantial increase in mineralization. Acted synergistically with BMP to form bone matrix.	Henstock et al. (2014)
TREK1 ion channel	Human MSC in mouse dorsum	Enhanced proteoglycan and collagen synthesis and extracellular matrix production and elevated the expression of type-1 and type-2.	Kanczler et al. (2010)
PDGFR1	Human MSC	Increased mineral/matrix ratio. Upregulation of osteogenic markers.	Hu et al. (2013)
Laminin-I	C2C12 myoblasts	Induced integrin activation in C2C12 cells, enhanced FAK phosphorylation and increased expression of myoblast differentiation markers, including galectin-1, α-enolase, and annexin III.	Grossi et al. (2007)
EGF	MTLn3 cells adenocarcinoma	EGF receptor activation including actin polymerization and ERK activation.	Kempiak and Segall (2004)
Integrin (RGD)	Neurites	Enhanced endothelin-1 gene expression by more than 100%.	Chen et al. (2001)
Integrin (antiβ1-Ab)	Neurites	Neurite elongation.	Chen et al. (2001)

Note: By exerting fine control over the local forces acting on mechanically sensitive proteins and structures, an element of external control is possible over these complex processes.

cell manipulation strategies provide potential therapeutic applications in cell therapies for soft and hard tissue repair (Figure 9.4).

The authors have previously reported success in remote magnetic activation of several targets, which in combination with soluble osteogenic factors in the culture media promoted the osteogenic differentiation of stem cells and the generation of denser, more abundant and calcified extracellular matrix in a 3D hydrogel scaffold (Henstock et al., 2014) (Figure 9.5). Using osteogenic media in combination with mechanical stimulation of the stem cells resulted in bone tissue being formed, a response that was further enhanced by adding BMPs to the culture (Henstock et al., 2014). The interaction of growth factor regulation of tissue formation with mechanical stimuli is a rapidly developing field of study which may have significant applications in both

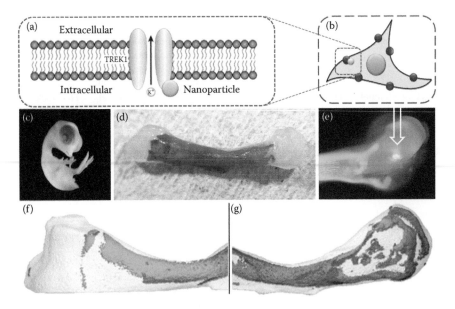

FIGURE 9.4

Magnetic nanoparticle activation of receptors and ion channels provides a mechanism for remote control of injected stem cells. As a route toward deliverable cell therapies, remote mechanoactivation has been validated by the authors using an ex vivo organotypic chick fetal femur model of endochondral ossification. Expansion and labeling of the TREK ion channel was performed during conventional monolayer cell culture (a, b). Chick femurs are isolated from embryonic day 11 fetuses (c) and cultured organotypically *in vitro* for 2 weeks (d). Microinjection of the labeled cells was *via* a needle into the unmineralized epiphysis, shown here by live-cell labeling with cell-tracker dye, DiO (e). After 14 days with daily magnetic stimulation using the bioreactor, a comparison of sham injected controls (f) and femurs microinjected with TREK nanoparticles (g) by x-ray microtomography revealed significantly more mineralization in the nanoparticles group, with up to 30% more bone being formed. Mineralization of the chick femur is by endochondral ossification—replacement of a cartilaginous precursor scaffold with bone, a developmental process that is often emulated as a strategy in regenerative medicine approaches for restoring bone defects.

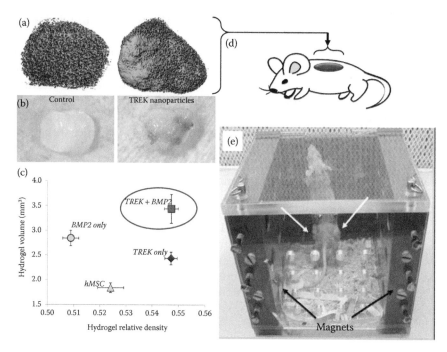

FIGURE 9.5

In addition to monolayer cultures of cells on tissue culture plastic substrates, research has shown that the MICA nanoparticle technology can be applied to cells cultured in 3D scaffolds. The approach is particularly useful in delivering mechanical stimuli to cells seeded within highly biocompatible hydrogel scaffolds that are too soft to endure direct physical loading using conventional bioreactors. In a study reported in *Stem Cells Translational Medicine* (2014), the authors cultured human MSCs labeled with anti-TREK MNPs in type I collagen hydrogels for 4 weeks with intermittent activation of the ion channel using the magnetic bioreactor (From Henstock, J. R. et al., *Stem Cells Transl Med*, 3, 1363–1374, 2014). MICA-stimulated hydrogels were shown to mineralize significantly more than unlabeled controls as determined by both x-ray microtomography of the mineralized phase within each hydrogel **(a)** and alizarin red staining for calcium deposition by differentiated osteoblasts **(b)**. TREK1 activation was shown to result in increased density of the mineralized nodules, a clinical marker of improved bone quality and function **(c)**. A parallel study incorporating bone morphogenetic protein (BMP2)-releasing microspheres revealed that the typical effects of BMP2 supplementation (poor-quality over-production of mineralized matrix) could be converted into a synergistic increase in both bone quantity and quality by appropriate mechanostransduction resulting from nanoparticle acti-vation of the TREK1 ion channel, as determined by microcomputed tomography (μCT). These hydrogels also can be introduced subcutaneously into mice to monitor their behavior *in vivo* **(d)**. In a previous study, Kanczler et al. (2010) developed a method for delivering variable mag-netic fields to mice without having to immobilize the animal. A dynamic magnetization of the SPIO nanoparticles was achieved by allowing the implanted mice to move freely between two opposing magnetic plates, resulting in minimal trauma to the experimental animals (From Kanczler, J. M. et al., *Tissue Eng Part A*, 16, 3241–3250, 2010)**(e)**.

developmental biology and tissue engineering (Kopf et al., 2012; Schwarz et al., 2013; Nowlan et al., 2010).

One particular advantage of this mechanoactivation technology is that is does not rely on the mechanical properties of the biomaterial scaffold, allowing cells in soft, tissue-mimicking hydrogel matrices to be mechanically stimulated. A new generation of highly-biocompatible, porous or hydrogel materials with active surfaces and structures has yielded a wealth of interesting, adaptable materials for applications in regenerative medicine that are handicapped only by their poor mechanical properties (Moreira Teixeira et al., 2014; Johnson and Christman, 2013). When implanted into load-bearing areas of the body these materials would rapidly fail, while in mechanically unstimulated defects the introduced cells do not receive the stimulatory loading required to generate optimum bone matrix (Ahearne, 2014). Using MNPs, loading can be applied directly to mechanoreceptors without deforming or loading the scaffold, stimulating the process of cell differentiation and the production of stronger, load-bearing extracellular matrix (Henstock et al., 2014). This opens the possibility for creating cell-seeded materials that can be implanted, injected, or molded into bone defects and which can now be mechanically stimulated to provide the osteogenic cues cells need for optimum tissue formation.

9.8 Conclusion

Tissue-engineered and regenerative therapies for medicine are now beginning to become clinical reality, with numerous ongoing trials in human patients that have primarily focused on establishing safety parameters for cell-based therapies. The next generation of tissue-engineered regenerative medicines will need to focus on efficacy, and optimizing these processes will inevitably require a knowledge and exploitation of all aspects of the therapy, from cell source, biomaterial scaffold, associated pharmacology, and of course, mechanics.

Having a remotely controllable mechanotransduction switch that can regulate aspects of healing from outside the body could be of enormous advantage to clinicians, allowing the process of tissue regeneration to be both optimized and controlled. By providing a wide choice of targets and associated complex signaling pathways, functionalized MNPs offer a unique range of possibilities for nonsurgical interventions in optimized cell therapies.

The biocompatible surface coating of MNPs can be functionalized with ligands or antibodies that directly bind the nanoparticle to a chosen receptor target, enabling direct control of the receptor's activity and downstream signal transduction by simply modulating the surrounding magnetic field.

The progress of these nanomagnetic actuation strategies is currently restricted by a poor understanding of the fundamental mechanisms that underlie how mechanical signals interact with biochemical factors and are integrated into the total cell response to a combination of complex stimuli. There is also some way to go in designing a "perfect" magnetic nanoparticle and in realizing the translational pathway for benchtop bioreactors to scalable bedside magnetic actuators. However, the potential for this technology is enticing.

References

Ahearne, M. (2014). Introduction to cell-hydrogel mechanosensing. *Interface Focus* 4:20130038.

Albro, M.B., et al. (2012). Shearing of synovial fluid activates latent TGF-β. *Osteoarthritis Cartilage* 20(11):1374–82.

Cartmell, S.H., et al. (2002). Development of magnetic particle techniques for long-term culture of bone cells with intermittent mechanical activation. *IEEE Trans Nanobioscience* 1(2):92–7.

Cartmell, S.H., et al. (2005). Use of magnetic particles to apply mechanical forces for bone tissue engineering purposes. *J Phys* 17:77–80.

Chemin, J., et al. (2005). A phospholipid sensor controls mechanogating of the K+ channel TREK-1. *EMBO J* 24:44–53.

Chen, J.X., et al. (2001). Twisting integrin receptors increases endothelin-1 gene expression in endothelial cells. *Am J Physiol Cell Physiol* 280(6):C1475–84.

Colombo, M., et al. (2012). Biological applications of magnetic nanoparticles. *Chem Soc Rev* 41(11):4306–34.

Creixell, M., Bohórquez, A.C., Torres-Lugo, M., Rinaldi, C. (2011). EGFR-targeted magnetic nanoparticle heaters kill cancer cells without a perceptible temperature rise. *ACS Nano* 5(9):7124–9. doi:10.1021/nn201822b.

Daquinag, A.C., Souza, G.R., Kolonin, M.G. (2013). Adipose tissue engineering in three-dimensional levitation tissue culture system based on magnetic nanoparticles. *Tissue Eng Part C Methods* 19(5):336–44.

Dobson, J. (2008). Remote control of cellular behaviour with magnetic nanoparticles. *Nat Nanotechnol* 3(3):139–43.

Dobson, J., et al. (2006). Principles and design of a novel magnetic force mechanical conditioning bioreactor for tissue engineering, stem cell conditioning, and dynamic *in vitro* screening. *IEEE Trans Nanobioscience* 5(3):173–7.

Dobson. J., St Pierre, T. (1996). Application of the ferromagnetic transduction model to D.C. and pulsed magnetic fields: Effects on epileptogenic tissue and implications for cellular phone safety. *Biochem Biophys Res Commun* 227(3):718–23.

El Haj, A.J., et al. (2015). An *in vitro* model of mesenchymal stem cell targeting using magnetic particle labelling. *J Tissue Eng Regen Med* 9:724–33.

Evans, E.A., Calderwood, D.A. (2007). Forces and bond dynamics in cell adhesion. *Science* 316(5828):1148–53.

Fass, J.N., Odde, D.J. (2003). Tensile force-dependent neurite elicitation via anti-beta1 integrin antibody-coated magnetic beads. *Biophys J* 85(1):623–36.

Glossop, J.R., Cartmell, S.H. (2010). Tensile strain and magnetic particle force application do not induce MAP3K8 and IL-1B differential gene expression in a similar manner to fluid shear stress in human mesenchymal stem cells. *J Tissue Eng Regen Med* 4(7):577–9.

Grossi, A., Yadav, K., Lawson, M.A. (2007). Mechanical stimuli on C2C12 myoblasts affect myoblast differentiation, focal adhesion kinase phosphorylation and galectin-1 expression: A proteomic approach. *Cell Biol Int* 35(6):579–86.

Harrison, R., et al. (2016). Autonomous magnetic labelling of functional mesenchymal stem cells for improved traceability and spatial control in cell therapy applications. *J Tissue Eng Regen Med*. doi:10.1002/term.2133. [Epub ahead of print].

Henstock, J.R., El Haj, A.J. (2015). Controlled mechanotransduction in therapeutic MSCs: Can remotely controlled magnetic nanoparticles regenerate bones? *Regen Med* 10:377–80.

Henstock, J.R., et al. (2014). Remotely activated mechanotransduction via magnetic nanoparticles promotes mineralisation synergistically with BMP2: Applications for injectable cell therapy. *Stem Cells Transl Med* 3:1–12.

Hofmann, A., et al. (2009). Combined targeting of lentiviral vectors and positioning of transduced cells by magnetic nanoparticles. *Proc Natl Acad Sci U S A* 106(1):44–9.

Hu, B., Dobson, J., El Haj, A.J. (2014). Control of smooth muscle α-actin (SMA) upregulation in HBMSCs using remote magnetic particle mechano-activation. *Nanomedicine* 10:45–55.

Huang, Z.Y., et al. (2010). Deep magnetic capture of magnetically loaded cells for spatially targeted therapeutics. *Biomaterials* 31(8):2130–40.

Huang, et al., Deep magnetic capture of magnetically loaded cells for spatially targeted therapeutics. Biomaterials, 2010. 31(8): p. 2130-2140.

Hughes, S., Dobson, J., El Haj, A.J. (2007). Magnetic targeting of mechanosensors in bone cells for tissue engineering applications. *J Biomech* 40:S96–104.

Hughes, S., El Haj, A.J., Dobson, J. (2005a). Magnetic micro-and nanoparticle mediated activation of mechanosensitive ion channels. *Med Eng Phys* 27(9):754–62.

Hughes, S., et al. (2005b). The influence of static magnetic fields on mechanosensitive ion channel activity in artificial liposomes. *Eur Biophys J* 34(5):461–8.

Hughes, S., et al. (2006). Expression of the mechanosensitive 2PK+ channel TREK-1 in human osteoblasts. *J Cell Physiol* 206:738–48.

Hung, C.T., et al. (2004). A paradigm for functional tissue engineering of articular cartilage via applied physiologic deformational loading. *Ann Biomed Eng* 32(1):35–49.

Ito, A., et al. (2005). Construction and delivery of tissue-engineered human retinal pigment epithelial cell sheets, using magnetite nanoparticles and magnetic force. *Tissue Eng* 11(3–4):489–96.

Ito, A., et al. (2007). Construction of heterotypic cell sheets by magnetic force-based 3-D coculture of HepG2 and NIH3T3 cells. *J Biosci Bioeng* 104(5):371–8.

Ito, A., Honda, H., Kamihira, M. (2008). Construction of 3D tissue-like structure using functional magnetite nanoparticles. *Yakugaku Zasshi* 128(1):21–8.

Janmey, P.A., Miller, R.T. (2011). Mechanisms of mechanical signaling in development and disease. *J Cell Sci* 124(1):9–18.

Johnson, T.D., Christman, K.L. (2013). Injectable hydrogel therapies and their delivery strategies for treating myocardial infarction. *Expert Opin Drug Deliv* 10:59–72.

Jurchenko, C., Salaita, K.S. (2015). Lighting up the force: Investigating mechanisms of mechanotransduction using fluorescent tension probes. *Mol Cell Biol* 35(15):2570–82.

Kanczler, J.M., et al. (2010). Controlled differentiation of human bone marrow stromal cells using magnetic nanoparticle technology. *Tissue Eng Part A* 16(10):3241–50.

Kempiak, S.J., Segall, J.E. (2004). Stimulation of cells using EGF-coated magnetic beads. *Sci STKE* 2004(218):pl1.

Khan, K.M., Scott, A. (2009). Mechanotherapy: How physical therapists' prescription of exercise promotes tissue repair. *Br J Sports Med* 43(4):247–52.

Kilgus, C., et al. (2012). Local gene targeting and cell positioning using magnetic nanoparticles and magnetic tips: Comparison of mathematical simulations with experiments. *Pharm Res* 29(5):1380–91.

Kirkham, G.R., et al. (2010). Hyperpolarization of human mesenchymal stem cells in response to magnetic force. *IEEE Trans Nanobioscience* 9(1):71–4.

Kirschvink, J.L., Kobayashi-Kirschvink, A., Woodford, B.J. (1992). Magnetite biomineralization in the human brain. *Proc Natl Acad Sci U S A* 89(16):7683–7.

Kobayashi, T., et al. (2008). A novel cell delivery system using magnetically labelled mesenchymal stem cells and an external magnetic device for clinical cartilage repair. *Arthroscopy* 24:69–76.

Kopf, J., Petersen, A., Duda, G.N., Knaus, P. (2012). BMP2 and mechanical loading cooperatively regulate immediate early signalling events in the BMP pathway. *BMC Biol* 10:37.

Kurayoshi, M., Yamamoto, H., Izumi, S., Kikuchi, A. (2007). Post-translational palmitoylation and glycosylation of Wnt-5a are necessary for its signalling. *Biochem J* 402:515–23.

Kyrtatos, P.G., et al. (2009). Magnetic tagging increases delivery of circulating progenitors in vascular injury. *JACC Cardiovasc Interv* 2(8):794–802.

Kwan and El Haj. (2014).

Laeremans, H., et al. (2011). Blocking of frizzled signaling with a homologous peptide fragment of wnt3a/wnt5a reduces infarct expansion and prevents the development of heart failure after myocardial infarction. *Circulation* 124:1626–35.

Lee, J-H., et al. (2006). Dual-mode nanoparticle probes for high-performance magnetic resonance and fluorescence imaging of neuroblastoma. *Angew Chem* 118:8340–2.

Lee, W.R., et al. (2011). Magnetic levitating polymeric nano/microparticular substrates for three-dimensional tumor cell culture. *Colloids Surf B Biointerfaces* 85(2):379–84.

Li, Q.Y., et al. (2013). Silica-coated superparamagnetic iron oxide nanoparticles targeting of EPCs in ischemic brain injury. *Biomaterials* 34(21):4982–92.

Lin, G.L., Hankenson, K.D. (2011). Integration of BMP, Wnt, and notch signaling pathways in osteoblast differentiation. *J Cell Biochem* 112:3491–501.

Lin, Z., et al. (2013). Thermoresponsive hydrogels from phosphorylated ABA triblock copolymers: A potential scaffold for bone tissue engineering. *Biomacromolecules* 14:2206–14.

Logan, C.W., Nusse, R. (2004). The Wnt signaling pathway in development and disease. *Cell Dev Bio* 20:781–810.

Luciani, A., et al. (2009). Magnetic targeting of iron-oxide-labeled fluorescent hepatoma cells to the liver. *Eur Radiol* 19(5):1087–96.

Mammoto, A.T., Ingber, D.E. Mechanosensitive mechanisms in transcriptional regulation. Journal of Cell Science, 2012. 125(13): p. 3061-3073.

Mannix, R.J., et al. (2008). Nanomagnetic actuation of receptor-mediated signal transduction. *Nat Nanotechnol* 3(1):36–40.

Markides, H., Kehoe, O., Morris, R.H., El Haj, A.J. (2013). Whole body tracking of superparamagnetic iron oxide nanoparticle-labelled cells—A rheumatoid arthritis mouse model. *Stem Cell Res Ther* 4:126.

Merkel, R., et al. (1999). Energy landscapes of receptor-ligand bonds explored with dynamic force spectroscopy. *Nature* 397(6714):50–3.

Miyazono, K. (2008). Shear activates platelet-derived latent TGF-beta. *Blood* 112(9):3533–4.

Mobasheri, A., et al. (2012). Potassium channels in articular chondrocytes. *Channels (Austin)* 6(6):416–25.

Moreira, Teixeira, L.S., Patterson, J., Luyten, F.P. (2014). Skeletal tissue regeneration: Where can hydrogels play a role? *Int Orthop* 38:1861–76.

Muthana, M., Kennerley, A.J., Hughes, R., Fagnano, E., Richardson, J., Paul, M., Murdoch, C., Wright, F., Payne, C., Lythgoe, M.F., Farrow, N., Dobson, J., Conner, J., Wild, J.M., Lewis, C. Directing cell therapy to anatomic target sites *in vivo* with magnetic resonance targeting. Nat Commun. 2015 Aug 18;6:8009

Na, S., et al. (2008). Rapid signal transduction in living cells is a unique feature of mechanotransduction. *Proc Natl Acad Sci U S A* 105(18):6626–31.

Nilius, B., Honoré, E. (2012). Sensing pressure with ion channels. *Trends Neurosci* 35(8):477–86.

Nowlan, N.C., et al. (2010). Mechanobiology of embryonic skeletal development: Insights from animal models. *Birth Defects Res C Embryo Today* 90:203–13.

Paluch, E.K., et al. (2015). Mechanotransduction: Use the force(s). *BMC Biology* 13:47.

Pankhurst, Q.A., et al. (2009). Progress in applications of magnetic nanoparticles in biomedicine. *J Phys D Appl Phys* 42(22).

Papachroni, K.K., et al. (2009). Mechanotransduction in osteoblast regulation and bone disease. *Trends Mol Med* 15(5):208–16.

Pankhurst, Q.A., et al., Progress in applications of magnetic nanoparticles in biomedicine. Journal of Physics D-Applied Physics, 2009. 42(22).

Pašukonienė, V., et al. (2014). Accumulation and biological effects of cobalt ferrite nanoparticles in human pancreatic and ovarian cancer cells. *Medicina* 50:237–244.

Perea, H., et al. (2006). Direct magnetic tubular cell seeding: A novel approach for vascular tissue engineering. *Cells Tissues Organs* 183(3):156–65.

Perea, H., et al. (2007). Vascular tissue engineering with magnetic nanoparticles: Seeing deeper. *J Tissue Eng Regen Med* 1(4):318–21.

Pereira, M., et al. (2007). Engineered cell-adhesive nanoparticles nucleate extracellular matrix assembly. *Tissue Eng* 13(3):567–78.

Pislaru, SV., et al. (2006). Magnetic forces enable rapid endothelialization of synthetic vascular grafts. *Circulation* 114:I314–8.

Polte, T.R., et al. (2007). Nanostructured magnetizable materials that switch cells between life and death. *Biomaterials* 28(17):2783–90.

Riegler, J., et al. (2013). Superparamagnetic iron oxide nanoparticle targeting of MSCs in vascular injury. *Biomaterials* 34(8):1987–94.

Riegler, J., et al., Superparamagnetic iron oxide nanoparticle targeting of MSCs in vascular injury. Biomaterials, 2013. 34(8): p. 1987-1994.

Rotherham, M., El Haj, A.J. (2015). Remote activation of the Wnt/β-catenin signalling pathway using functionalised magnetic particles. *PLOS ONE* 10:e0121761.

Muthana, M., Kennerley, A.J., Hughes, R., Fagnano, E., Richardson, J., Paul, M., Murdoch, C., Wright, F., Payne, C., Lythgoe, M.F., Farrow, N., Dobson, J., Conner, J., Wild, J.M., Lewis, C. Directing cell therapy to anatomic target sites *in vivo* with magnetic resonance targeting. Nat Commun. 2015 Aug 18;6:8009

Sawada, Y., et al. (2006). Force sensing by mechanical extension of the Src family kinase substrate p130Cas. *Cell* 127(5):1015–26.

Schwarz, C., et al. (2013). Mechanical load modulates the stimulatory effect of BMP2 in a rat non-union model. *Tissue Eng Part A* 19:247–54.

Shemesh, T., et al. (2005). Focal adhesions as mechanosensors: A physical mechanism. *Proc Natl Acad Sci U S A* 102(35):12383–8.

Shimizu, K., Ito, A., Honda, H. (2007b). Mag-seeding of rat bone marrow stromal cells into porous hydroxyapatite scaffolds for bone tissue engineering. *J Biosci Bioeng* 104(3):171–7.

Souza, G.R., et al. (2010). Three-dimensional tissue culture based on magnetic cell levitation. *Nat Nanotechnol* 5(4):291–6.

Storch, U., Schnitzler, M.M.Y., Gudermann, T. (2012). G protein-mediated stretch reception. *Am J Physiol Heart Circ Physiol* 302(6):H1241–9.

Tanase, M., et al. (2005). Assembly of multicellular constructs and microarrays of cells using magnetic nanowires. *Lab Chip* 5(6):598–605.

Tseng, H., et al. (2013). Assembly of a three-dimensional multitype bronchiole coculture model using magnetic levitation. *Tissue Eng Part C Methods* 19(9):665–75.

Vogel, V., Sheetz, M.P. (2009). Cell fate regulation by coupling mechanical cycles to biochemical signaling pathways. *Curr Opin Cell Biol* 21(1):38–46.

Wilhelm, C., et al. (2007). Magnetic control of vascular network formation with magnetically labeled endothelial progenitor cells. *Biomaterials* 28(26):3797–806.

Wimpenny, I., Markides, H., El Haj, A.J. (2012). Orthopaedic applications of nanoparticle-based stem cell therapies. *Stem Cell Res Ther* 3:13.

Xu, F., et al. (2011). Three-dimensional magnetic assembly of microscale hydrogels. *Adv Mater* 23(37):4254–60.

Xu, F., et al. (2012). Release of magnetic nanoparticles from cell-encapsulating biodegradable nanobiomaterials. *ACS Nano* 6(8):6640–9.

Yang, Q.Z., et al. (2009). A molecular force probe. *Nat Nanotechnol* 4(5):302–6.

10

Magnetic Nanoparticles for 3D Cell Culture

Hubert Tseng, Robert M. Raphael,
Thomas C. Killian, and Glauco R. Souza

CONTENTS

10.1 Introduction 229
10.2 Magnetic Cell Culture 231
10.3 Models 236
    10.3.1 Glioblastoma 236
    10.3.2 Adipose Tissue 238
    10.3.3 Bronchiole 238
    10.3.4 Aortic Valve 239
    10.3.5 Wound Healing Assay 239
10.4 Future Directions 240
References 241

10.1 Introduction

Amidst the ongoing nanotechnology revolution in biomedical research and therapeutic applications, magnetic nanotechnology has attracted increasing attention. The most traditional application of magnetic nanoparticles in biomedical research has been in imaging, where gadolinium (Gd) or iron oxide-based nanoparticles are commonly used as contrast agents for magnetic resonance imaging (MRI).[1–3] More recently, the application of magnetic nanoparticles has also extended to flow cytometry and magnetically assisted cell sorting (MACS), where antibody-conjugated nanoparticles target specific cell surface proteins to preferentially select specific cell populations with magnetic forces.[4] In these and other applications, nanoparticles have begun to replace microparticles, because of their increased specificity binding to target proteins and optical qualities.[5,6] Other more recent applications of magnetic nanoparticles utilize their magnetism to apply specific forces and direct cells. Although the toxicity of nanoparticles, including magnetic nanoparticles, is still a significant concern to their use, particularly clinical use,[7] the potential applications of magnetic nanoparticles could strongly impact biology and medicine in the near future.

This chapter focuses on the use of magnetic nanoparticles for three-dimensional (3D) cell culture. The principle behind this application is the magnetization of cells with magnetic nanoparticles.[8] Once magnetized, these cells can be manipulated and directed with the application of an external magnetic field. More specifically, magnetized cells can either be levitated (magnetic levitation) or printed on a surface (magnetic 3D bioprinting). When levitated or bioprinted, the magnetic forces attract and aggregate the cells, which begin to interact with each other, inducing the synthesis of extracellular matrix (ECM) and the formation of a larger 3D construct.[8–19]

The use of magnetic nanoparticles and fields for 3D cell culture holds several advantages over existing models for biomedical research (Figure 10.1). In a broader context, 3D cell culture is more representative of native tissue, and thus more predictive of human *in vivo* response, than traditionally used two-dimensional (2D) models.[20] Whereas cells in 2D are cultured on stiff plastic or glass substrates not found within the body, cells in 3D grow in a soft environment, either artificial or endogenously grown, that mimics the dimensionality, ECM, and cell–cell and cell–ECM interactions that exist in native tissue.[21–26] Moreover, cells in 2D are uniformly exposed to nutrients/biochemical factors in the media above it, while the exposure of cells in 3D varies based on its position within the structure and the composition of the surrounding ECM, making it less sensitive to compounds and exogenous factors and mimicking *in vivo* situations.[27,28] As a result, biomedical research is quickly gravitating toward 3D cell culture models that faithfully represent the native tissue.

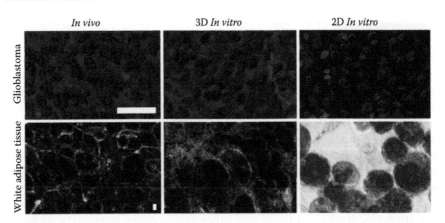

FIGURE 10.1

Comparison of 2D and 3D *in vitro* cell culture models to *in vivo* tissue. Immunohistochemical staining of (top) glioblastoma (N-cadherin = red) and (bottom) the SVF of WAT (perilipin = red, CD31 = green) *in vivo* (left), magnetically levitated in 3D (center), and culture in 2D (right). Nuclei were counterstained with 4′,6-diamidino-2-phenylindole (DAPI) (blue). Note the similarity in structure between the magnetically levitated 3D culture and the native tissue, as opposed to the cells in 2D. Scale bar = 25 μm. (Adapted from Souza, G. R. et al., *Nat. Nanotechnol.*, 5, 291–6, 2010; Daquinag, A. C. et al., *Tissue Eng. Part C Methods*, 19, 336–44, 2013. With permission.)

There is a wide variety of systems for 3D cell culture. Systems involving protein gels (e.g., collagen,[29] Matrigel[30,31]) or polymer scaffolds (e.g., polyethylene glycol[32]) use artificial substrates and specialized media to encapsulate cells. These models can largely recreate specific niche environments, yet are limited by scalability to other tissues, high costs, and long fabrication processes. Spheroid systems, such as hanging drop spheroids,[33,34] round bottom plates,[35] or nanopatterned plates,[36,37] use similar principles of cell aggregation and self-assembly as magnetic cell culture, but suffer from long spheroid formation times (3–7 days) and the lack of spatial control. Magnetic cell culture also holds several practical advantages over other 3D cell culture systems. First, using magnetic forces to attract and aggregate cells accelerates spheroid formation to within 24 hours.[8] Second, magnetic forces offer control of the 3D culture, whether it be the ability to print cells into cultures of desired patterns,[16] create coculture models,[14,17,18] or design innovative assays.[8,9] Along the same lines, magnetic forces can also be used to hold down 3D cultures while exchanging media or staining, thereby improving sample retention.[14,17,18] Finally, magnetic cell culture does not require any specialized media or equipment. Overall, magnetic nanoparticles can be applied to cell culture to rapidly and easily build 3D cell culture with fine spatial organization.

10.2 Magnetic Cell Culture

The basic principle underpinning magnetic cell culture is the magnetization of cells. Our group uses a magnetic nanoparticle assembly consisting of iron oxide, gold, and poly-L-lysine (PLL) that electrostatically attaches to the cell membrane via PLL.[8] This magnetization is nonspecific to cell type and does not require any specialized media or equipment to magnetize the cells. A previous version of these nanoparticles used M13 bacteriophage instead of PLL to enter the cell,[8] while other groups have used other magnetic particles to magnetize cells, including bacteria-derived magnetic nanoparticles,[38] and microparticles including iron oxide containing fibroin microspheres[39] and streptavidin paramagnetic particles.[40]

Given the prevalence of toxicity issues with nanoparticles,[7] our group has rigorously tested and consistently confirmed the cytocompatibility and bioinertness of our nanoparticle assembly. No effect of the nanoparticles has been found on cell proliferation, viability, metabolism, and inflammatory stress as measured by interleukins (IL) -6 and 8 in primary human lung cells[17] and porcine aortic valve cells.[18] In fact, glioblastoma (GBM)[8] and vascular smooth muscle cells (SMCs)[13] have been shown to grow faster in magnetic cell culture than in 2D. We have also found the nanoparticles to not interfere with most experimental techniques, as demonstrated by our and other groups' successes in performing the following experiments on magnetic 3D cultures: immunohistochemical and immunofluorescent staining,[8,14,15,17,18] confocal microscopy,[14] transmission electron microscopy (TEM),[8] Western blotting,[9,11] quantitative reverse transcriptase polymerase chain reaction (qRT-PCR),[18] and absorbance-based biochemical assays.[16,17] One experimental limitation to magnetic 3D culture is immunohisto-chemical staining with diaminobenzidine (DAB), which stains brown and can be easily confused with the brown nanoparticles.[15] All considered, the nanoparticles are cytocompatible and noninterfering.

Once magnetized, the cells are detached from their substrate, counted, resuspended in media, then distributed into a multiwell plate.[15] By externally applying a fixed magnetic field, the cells can be aggregated. The magnetic field strength needed to attract cells is mild (50–300 G), with no effect found proliferation, viability, metabolism, and inflammatory stress in primary human lung cells[17] and porcine aortic valve cells.[18] After the field is applied, with no stiff substrate to attach to, the cells will self-assemble into a 3D structure. This method of 3D cell culture is similar to others where cell aggregation and cell–cell adhesion are the fundamental drivers, like spheroids.[33–37] Yet, in magnetic cell culture, magnetic forces are used to accelerate cell aggregation from days to hours. Our studies investigating 3D cell culture for human primary lung cells has shown that within 6 hours of applying the magnetic field, laminin will be synthesized and extruded by the cells. By 48 hours, the 3D structure will have an organized ECM.[17]

The resulting 3D cell culture can then be maintained long term, up to a few months.[8] As the 3D cultures are magnetic, media can be exchanged without risk of losing cells by using magnets to hold the cultures down.[15] Similarly, 3D cultures can be held down for immunostaining.[15,17,18] These cultures can also be transferred easily using a magnetic pen consisting of a Teflon pen with a magnetic rod inside to pick up cultures and move them between vessels for coculture or imaging.[15,17,18]

Using these basic principles, we have developed several methods to create 3D cultures. The first method is magnetic levitation, in which a magnetic field is applied from above the culture vessel to levitate magnetized cells (Figure 10.2).[8] The cells will levitate and aggregate within the media, with the air–liquid interface serving as a physical barrier to contain the 3D culture within the media. The second method is magnetic 3D bioprinting, in which the magnetic field is applied below the well to rapidly attract cells to the bottom of an ultra-low attachment plate (Figure 10.3).[16] In this method, cells aggregate into a pattern defined by the magnet, and allow for the printing of 3D cellular aggregates into patterns of interest, such as rings or spheroids. The practical difference between the two methods is in size and reproducibility of the resulting cultures. While magnetic levitation can create large 3D cultures with high cell numbers and unconstrained growth,

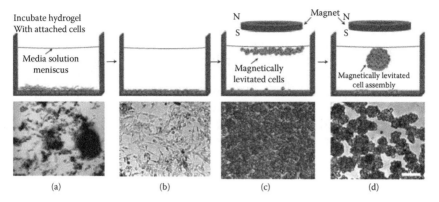

FIGURE 10.2
Schematic of magnetic levitation. (a) Cells grown in 2D are statically incubated overnight with the nanoparticles to magnetize them. (b) When magnetized, the cells will appear peppered with the nanoparticles. (c) The next day, the cells are detached from their substrate, resuspended in media, then distributed into a multiwell plate, where a magnet atop the plate is used to levitate the cells off the surface and into the media. (d) Over time, the cells will aggregate and interact with each other to form a larger 3D culture. Scale bar = 30 μm. (Reprinted from Souza, G. R. et al., *Nat. Nanotechnol.*, 5, 291–296, 2010. With permission.)

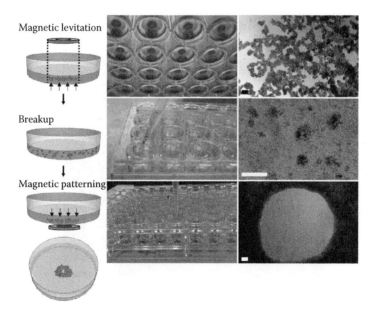

FIGURE 10.3
Schematic of magnetic 3D bioprinting. Cartoon (left) with corresponding images (center) and brightfield microscopy images of HEK293s in 3D (right). Magnetized cells are first levitated to induce ECM synthesis (top). Then, the cells are mechanically disrupted using pipette action (center), and patterned into ring shapes (bottom). Scale bar = 100 μm. (Reprinted from Timm, D. M. et al., *Sci. Rep.*, 3, 3000, 2013. With permission.)

magnetic 3D bioprinting prints smaller cultures in patterns determined by the magnet, making the method ideal for high-throughput 3D cell printing with reproducibility (Figure 10.4).

From the foundation of magnetic cell culture, we have developed models and assays from the simple to the complex to mimic native tissue responses. Given the nonspecific nature of the nanoparticles, our group has grown numerous cell types in 3D with little to no trouble (see Haisler et al. for a list of cells cultured in 3D using magnetic nanoparticles by our group).[15] One focus has been on coculture models that mimic the complex cellular environments found in native tissue. The simplest coculture can be made by mixing two cell types that are individually magnetized. We have previously created a coculture of adipocytes and endothelial cells by straight mixture and cultured it for 14 days in 3D (Figure 10.4).[14] While the cells in 2D lost their heterogeneity after several passages due to the fibroblasts taking over, in 3D, not only were the cells' phenotypes maintained, but the cells self-organized.[14] Our group has also designed layered cocultures in which cells of different types are separately cultured in 3D, then assembled into a layered coculture using a magnetic pen (Figure 10.5). This approach was used to mimic the layered structures of the lung[17] and the aortic valve (Figure 10.6).[18]

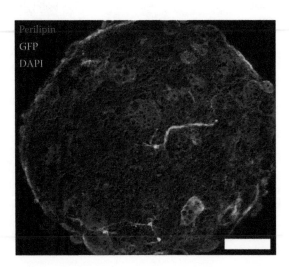

FIGURE 10.4
Immunohistochemical staining of a magnetically levitated 3D culture of 3T3-L1 murine preadipocytes and GFP-transfected bEND.3 murine endothelial cells (green) for perilipin (red) after 14 days. Nuclei were counterstained with DAPI (blue). Note the formation of multicellular lipid droplets and the vascular network formed by the endothelial cells, as seen *in vivo*. Scale bar = 50 μm. (Adapted from Daquinag, A. C. et al., *Tissue Eng. Part C Methods*, 19, 336–44, 2013. With permission.)

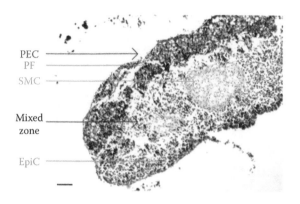

PEC
PF
SMC

Mixed
zone

EpiC

FIGURE 10.5
Hematoxylin & eosin (H&E) staining of a layered coculture model of the bronchiole using magnetic levitation after 2 days. Note the maintenance of a layered structure and organization between the cells. In particular, PFs formed a collagenous layer, while SMCs formed contractile regions similar to that seen *in vivo*. Scale bar = 100 μm. (Reprinted from Tseng, H. et al., *Tissue Eng. Part C Methods*, 19, 665–75, 2013. With permission.)

We have also designed assays to mimic specific situations. Our first assay was an invasion assay, wherein two separate 3D cultures of different cell types, one cancer type and one normal type, are transferred into the same well, levitated together, and the time for the cancer cell to take over the culture is recorded. We demonstrated that GBM take over normal human astrocytes (NHAs) over 10 days.[8,9] More recently, we used magnetic 3D bioprinting to print cells into 3D rings that we found to close soon after printing, and do so in a drug- and dose-dependent manner.[16] Ring closure mimics the mechanism of wound healing assays, where a cell monolayer's ability to close a wound is measured, which correlates with cell proliferation and migration.[41–44] This assay recapitulates wound healing, but in a representative 3D environment and an experimental platform amenable to high-throughput toxicity screening (Figure 10.7). We have also magnetically bioprinted more complex patterns, such as the letter N using green fluorescent protein (GFP) labeled GBM cells (Figure 10.8).

Taken together, magnetic nanoparticles are useful tools for 3D culture. Magnetic forces are used to rapidly aggregate magnetized cells. These cells self-assemble into larger 3D cultures with ECM that mimics native tissue. The nanoparticles and magnetic fields are biocompatible and noninterfering. The magnetism of the cultures facilitates basic functions, such as exchanging media and staining, but also permits the design of more complex models for research. In the following section, we discuss the specific cellular models our group and others have designed.

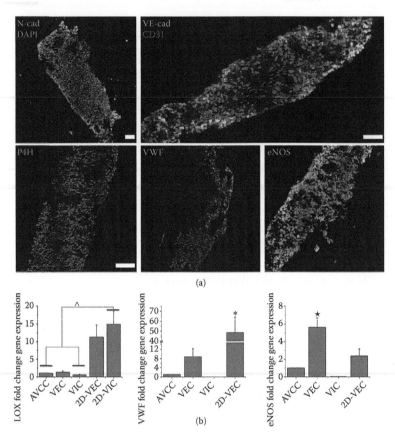

FIGURE 10.6
(a) IHC stains (green) for the functional markers N-cad, P4H, von Willebrand factor (VWF), eNOS, and VE-cad (double-stained with CD31 in red) in the aortic valve coculture (AVCC). Nuclei are counterstained using DAPI (blue). The endothelial markers VE-cad and VWF were localized to the outer edges of the AVCC. Scale bar = 100 μm. (b) qRT-PCR results for the functional markers LOX, VWF, and eNOS (n = 3–5). Significant reduction in eNOS expression in the AVCC compared to VECs in 2D and 3D suggests their transdifferentiation. *p < .05 vs. other groups. ^p < .05 within bracket. Error bars represent standard error of the mean. (Reprinted from Tseng, H. et al., *Acta Biomater*, 10, 173–82, 2014. With permission.)

10.3 Models

10.3.1 Glioblastoma

Our first reported model using magnetic cell culture was an invasion assay to assess the invasiveness of GBM cells.[8,9] mCherry-transfected NHAs and GFP-transfected human GBM were magnetically levitated and cultured separately for 3–5 days. Once both cultures were of the same size, they were then transferred to the same well and relevitated, where the magnetic field

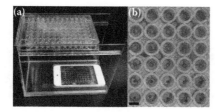

FIGURE 10.7
(a) The mobile device-based imaging setup to image rings for wound healing in 3D. A 96-well plate full of rings is placed on the top of the setup. At the bottom of the setup sits the mobile device with the camera facing upwards to image the whole plate. By imaging whole plates, the system forgoes costly and time-consuming well-by-well imaging under a microscope. (b) A sample image taken with the mobile device of 30 rings of HEK293s dosed with ibuprofen. Note the dark color and the resolution of the rings within the media. Scale bar = 5 mm.

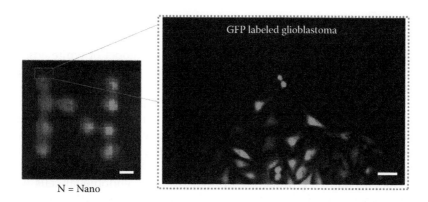

FIGURE 10.8
GFP labeled glioblastoma cells (LN229) magnetically bioprinted in the shape of the letter N (left image, scale bar = 5 mm) on the bottom of a standard tissue culture plate. Right image, cells from the top left of the left image magnified showing individual cells (right image, scale bar = 25 μm).

brought the two cultures in contact with each other. From here, the GBM could invade the NHA, a process that could be tracked using fluorescent microscopy (Figure 10.9). Our group demonstrated that a GBM cell line (LN229) could fully take over the NHAs in 3D by 10.5 days.[8] This assay was then used by another group to assess differences in invasiveness between subpopulations in GBM tumors, specifically between core cells and invasive cells. Their results demonstrated that at 2 days, invasive cells began to invade the NHA culture, while core cells had not. This difference in invasiveness between the subpopulations was correlated with the acquisition of stem cell markers and increased Akt activation in invasive cells.[9] Overall,

12 h	48 h	72 h	156 h	252 h

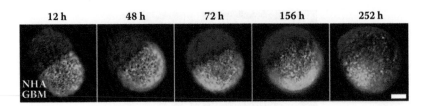

FIGURE 10.9
Invasion assay with NHAs (red) and GBMs (green). NHAs and GBMs were levitated in separate wells for 3–5 days. Afterward, they were transferred to the same well, relevitated under the same magnetic field, and GBMs were allowed to invade the NHAs. GBMs were found to take 10.5 days to take over the NHAs. Scale bar = 200 µm. (Adapted from Souza, G. R. et al., *Nat. Nanotechnol.*, 5, 291–6, 2010. With permission.)

we demonstrated the capability of magnetic cell culture to design unique 3D assays that mimic *in vivo* situations.

10.3.2 Adipose Tissue

In this project, we sought to build a 3D coculture model of white adipose tissue (WAT), which contains a heterogeneous cell population consisting of adipocytes and the stromal vascular fraction (SVF) of adipose stromal cells (ASCs)/adipocyte progenitors, vascular endothelial cells (VECs), and infiltrating leukocytes.[45,46] When grown in 2D, SVF cells lose their heterogeneity, as endothelial cells and hematopoietic cells are lost at later passages as mesenchymal cells take over,[46–48] necessitating the need for a 3D coculture model for *in vitro* testing. As a proof of concept, 3T3-L1 murine embryonic preadipocytes and GFP-transfected bEND.3 murine endothelial cells were separately magnetized, then mixed and levitated together. After 14 days of adipogenesis, these adipospheres revealed an abundance of perilipin droplets, a mature lipid marker, and an endothelial network (Figure 10.4).[14] Based off this success, a mixture of SVF cells derived from mouse WAT was levitated, and adipogenesis was induced. The adiposphere maintained heterogeneity in phenotype, with immunohistochemical staining for CD31 and decorin confirming the presence of vascular lumens and ECM within the adiposphere.[14] By 14 days, the adiposphere closely resembled the original mouse WAT in structure. This model faithfully represents the native WAT *in vitro*, thus overcoming the shortcomings of 2D culture.

10.3.3 Bronchiole

To demonstrate the ability to create layered cocultures, we built a model of the bronchiole from airway to circulation, which is natively layered of four cell types.[49] The four constituent cell types of the bronchiole, all human and primary, were all separately magnetically levitated: bronchial epithelial cells (EpiC); tracheal SMCs; pulmonary fibroblasts (PFs); and pulmonary endothelial cells (PECs).[17] After 4 hours of levitation, the coculture was assembled by using the magnetic pen to sequentially stack layers of cells; first EpiCs, then SMCs, PFs, and PECs. The coculture was held on the pen for 4 hours to encourage adhesion between the layers, before being released and relevitated. After 2 days, the coculture still maintained all of its constituent cell phenotypes, and induced the synthesis and organization of ECM, particularly collagen, to mimic the layered structure of the lung (Figure 10.5).[17] After 7 days, epithelial phenotype was still maintained with and without serum, demonstrating that neither serum nor long-term culture would result in the loss of epithelial phenotype.[17] The results of this study confirmed our ability to layered cocultures with fine spatial control that maintained phenotype and induced ECM synthesis.

10.3.4 Aortic Valve

We then applied the coculture technique to develop a coculture model of the aortic valve.[18] The aortic valve consists of aortic valve interstitial cells (VICs) and VECs. Both cell types and their interactions are indicated in the etiology of calcific aortic valve disease (CAVD), yet their exact roles are unclear given the paucity of successful coculture models. The few models that do exist were collagen-based models that required days to form.[50,51] In this magnetically levitated coculture model, both VICs and VECs were levitated for 4 hours, then sequentially assembled into a coculture using the MagPen. The cells were assembled with VECs first, then VICs. After 4 hours on the pen, the coculture was released and relevitated. Both phenotypes were preserved within the coculture after 2 days and ECM was synthesized and organized, all demonstrated by immunohistochemistry and qRT-PCR. Colocalization of CD31 and αSMA and a decrease in genetic expression of eNOS within the coculture suggested the undergoing of endothelial-mesenchymal transdifferentiation (EMT), which could potentially lead to the use of VECs as a cell source for heart valve tissue engineering (Figure 10.6). We also found through qRT-PCR that VECs help to regulate the contractile phenotype of VICs commonly associated with calcification, confirming previous reports of that role and CAVD progression as a result of VEC dysfunction.[18,50] Altogether, magnetic cell culture was applied to develop a 3D model of the aortic valve *in vitro* where others have been less successful.

10.3.5 Wound Healing Assay

Most recently, we used magnetic 3D bioprinting to develop wound healing assays in 3D for high-throughput toxicity testing.[16] In wound healing or scratch assays, monolayers of cells are injured to create a void.[41-44] Healthy cells are able to close the wound as a function of migration, and the amount of closure can be related to drug concentration to assess toxicity. However, these assays are performed in 2D environments that poorly represent native tissue. Thus, this study designed a 3D version of this assay using magnetic 3D bioprinting. Magnetized cells, either human embryonic kidney cells (HEK293) or SMCs, were first levitated overnight to induce ECM synthesis. The next day, the cells were disrupted, resuspended in media, distributed into a multiwell plate, then printed into rings using ring-shaped magnets. After holding the rings on the magnet for 1 hour, either ibuprofen or sodium dodecyl sulfate (SDS) was added to the rings in varying concentrations. When released off the magnet, the rings will close in a dose-dependent manner. Ring closure was tracked using a mobile device-based imaging system, which was possible given the large size of the rings (0.1875″ OD, 0.0625″ ID), the brown color of the magnetized cells, and the computing power of commercially available mobile devices. The mobile device was programmed using a freely available app (Experimental Assistant, Nano3D Biosciences) to image whole plates of rings at specific time intervals (Figure 10.7). These images were then batch analyzed afterward to measure ring closure over time, which was then used to find relevant metrics of toxicity, such as the half maximum inhibitory concentration (IC_{50}). By imaging whole plates at once, time-consuming imaging of each individual well under a microscope is avoided, improving throughput and efficiency. We found ring closure to be dose-dependent and correlate with viability and migration. These results demonstrated the application of magnetic cell culture for high-throughput compound screening in toxicity assays.

10.4 Future Directions

The versatility and benefits of magnetic nanoparticles for 3D cell culture leave several opportunities to pursue in the future. For example, we have established that the wound healing assay can be used as a general toxicity screen. This assay can be tailored for specific applications with minor modification, wherein the response rings or spheroids could be translated to an *in vivo* response, such as vasoactivity or tumor growth. Additionally, as screening libraries hold 50,000–300,000 compounds on average,[52] the direction of all 3D cell culture technologies will be toward improving throughput, and adapting magnetic cell culture, for these processes will be critical. Finally, with our magnetic nanoparticles being biocompatible and noninterfering with imaging, future opportunities lie in their use for therapeutics. We expect that further development in the applications of magnetic nanoparticles for cell culture will continue to introduce exciting new approaches for controlling size, shape, and composition of 3D tissue constructs, and have a large impact in life science research and medicine.

References

1. Sun, C., Lee, J. S. H., and Zhang, M. Magnetic nanoparticles in MR imaging and drug delivery. *Adv. Drug Deliv. Rev.* **60**, 1252–65 (2008).
2. Rozanova, N. and Zhang, J. Z. Metal and magnetic nanostructures for cancer detection, imaging, and therapy. *J. Biomed. Nanotechnol.* **4**, 377–99 (2008).
3. Pankhurst, Q. A., Connolly, J., Jones, S. K., and Dobson, J. Applications of magnetic nanoparticles in biomedicine. *J. Phys. D Appl. Phys.* **36**, R167–81 (2003).
4. Wedemeyer, N. and Pötter, T. Flow cytometry: An "old" tool for novel applications in medical genetics. *Clin. Genet.* **60**, 1–8 (2001).
5. Grützkau, A. and Radbruch, A. Small but mighty: How the MACS-technology based on nanosized superparamagnetic particles has helped to analyze the immune system within the last 20 years. *Cytometry. A* **77**, 643–7 (2010).
6. He, J., Huang, M., Wang, D., Zhang, Z., and Li, G. Magnetic separation techniques in sample preparation for biological analysis: A review. *J. Pharm. Biomed. Anal.* **101**, 84–101 (2014). doi:10.1016/j.jpba.2014.04.017.
7. Zhang, M., Jin, J., Chang, Y.-N., Chang, X., and Xing, G. Toxicological properties of nanomaterials. *J. Nanosci. Nanotechnol.* **14**, 717–29 (2014).
8. Souza, G. R. et al. Three-dimensional tissue culture based on magnetic cell levitation. *Nat. Nanotechnol.* **5**, 291–6 (2010).
9. Molina, J. R., Hayashi, Y., Stephens, C., and Georgescu, M. -M. Invasive glioblastoma cells acquire stemness and increased Akt activation. *Neoplasia* **12**, 453–63 (2010).
10. Lee, J. S., Morrisett, J. D., and Tung, C. -H. Detection of hydroxyapatite in calcified cardiovascular tissues. *Atherosclerosis* **224**, 340–7 (2012).
11. Xu, L., Gao, G., Ren, J., Su, F., and Zhang, W. Estrogen receptor β of host promotes the progression of lung cancer brain metastasis of an orthotopic mouse model. *J. Cancer Ther.* **3**, 352–8 (2012).
12. Becker, J. L. and Souza, G. R. Using space-based investigations to inform cancer research on Earth. *Nat. Rev. Cancer* **13**, 315–27 (2013).
13. Castro-Chavez, F., Vickers, K. C., Lee, J. S., Tung, C. -H., and Morrisett, J. D. Effect of lyso-phosphatidylcholine and Schnurri-3 on osteogenic transdifferentiation of vascular smooth muscle cells to calcifying vascular cells in 3D culture. *Biochim. Biophys. Acta* **1830**, 3828–34 (2013).
14. Daquinag, A. C., Souza, G. R., and Kolonin, M. G. Adipose tissue engineering in three-dimensional levitation tissue culture system based on magnetic nanoparticles. *Tissue Eng. Part C Methods* **19**, 336–44 (2013).
15. Haisler, W. L. et al. Three-dimensional cell culturing by magnetic levitation. *Nat. Protoc.* **8**, 1940–9 (2013).
16. Timm, D. M. et al. A high-throughput three-dimensional cell migration assay for toxicity screening with mobile device-based macroscopic image analysis. *Sci. Rep.* **3**, 3000 (2013).
17. Tseng, H. et al. Assembly of a three-dimensional multitype bronchiole coculture model using magnetic levitation. *Tissue Eng. Part C Methods* **19**, 665–75 (2013).
18. Tseng, H. et al. A three-dimensional co-culture model of the aortic valve using magnetic levitation. *Acta Biomater.* **10**, 173–82 (2014).
19. Marx, V. Cell culture: A better brew. *Nature* **496**, 253–8 (2013).

20. Abbott, A. Cell culture: Biology's new dimension. *Nature* **424**, 870–2 (2003).
21. Griffith, L. G. and Swartz, M. A. Capturing complex 3D tissue physiology *in vitro*. *Nat. Rev. Mol. Cell Biol.* **7**, 211–24 (2006).
22. Peyton, S. R., Kim, P. D., Ghajar, C. M., Seliktar, D., and Putnam, A. J. The effects of matrix stiffness and RhoA on the phenotypic plasticity of smooth muscle cells in a 3-D biosynthetic hydrogel system. *Biomaterials* **29**, 2597–607 (2008).
23. Pedersen, J. A. and Swartz, M. A. Mechanobiology in the third dimension. *Ann. Biomed. Eng.* **33**, 1469–90 (2005).
24. Cukierman, E., Pankov, R., Stevens, D. R., and Yamada, K. M. Taking cell-matrix adhesions to the third dimension. *Science* **294**, 1708–12 (2001).
25. Pampaloni, F., Reynaud, E. G., and Stelzer, E.H.K. The third dimension bridges the gap between cell culture and live tissue. *Nat. Rev. Mol. Cell Biol.* **8**, 839–45 (2007).
26. Kleinman, H. K., Philp, D., and Hoffman, M. P. Role of the extracellular matrix in morphogenesis. *Curr. Opin. Biotechnol.* **14**, 526–32 (2003).
27. Zhang, S. Beyond the Petri dish. *Nat. Biotechnol.* **22**, 151–2 (2004).
28. Ng, K. W., Leong, D. T. W., and Hutmacher, D. W. The challenge to measure cell proliferation in two and three dimensions. *Tissue Eng.* **11**, 182–91 (2005).
29. Brown, R. A. In the beginning there were soft collagen-cell gels: Towards better 3D connective tissue models? *Exp. Cell Res.* **319**, 2460–9 (2013).
30. Benton, G., Kleinman, H. K., George, J., and Arnaoutova, I. Multiple uses of basement membrane-like matrix (BME/Matrigel) *in vitro* and in vivo with cancer cells. *Int. J. Cancer* **128**, 1751–7 (2011).
31. Benton, G., Arnaoutova, I., George, J., Kleinman, H. K., and Koblinski, J. Matrigel: From discovery and ECM mimicry to assays and models for cancer research. *Adv. Drug Deliv. Rev.* **79** –80, 3–18 (2014). doi:10.1016/j.addr.2014.06.005.
32. DeVolder, R. and Kong, H. -J. Hydrogels for in vivo-like three-dimensional cellular studies. *Wiley Interdiscip. Rev. Syst. Biol. Med.* **4**, 351–65 (2012).
33. Messner, S., Agarkova, I., Moritz, W., and Kelm, J. M. Multi-cell type human liver microtissues for hepatotoxicity testing. *Arch. Toxicol.* **87**, 209–13 (2013).
34. Amann, A. et al. Development of an innovative 3D cell culture system to study tumour—Stroma interactions in non-small cell lung cancer cells. *PLOS ONE* **9**, e92511 (2014).
35. Ivascu, A. and Kubbies, M. Rapid generation of single-tumor spheroids for high-throughput cell function and toxicity analysis. *J. Biomol. Screen.* **11**, 922–32 (2006).
36. Yoshii, Y. et al. The use of nanoimprinted scaffolds as 3D culture models to facilitate spontaneous tumor cell migration and well-regulated spheroid formation. *Biomaterials* **32**, 6052–8 (2011).
37. Rimann, M. and Graf-Hausner, U. Synthetic 3D multicellular systems for drug development. *Curr. Opin. Biotechnol.* **23**, 803–9 (2012).
38. Kim, J. A. et al. High-throughput generation of spheroids using magnetic nanoparticles for three-dimensional cell culture. *Biomaterials* **34**, 8555–63 (2013).
39. Lee, J. H. and Hur, W. Scaffold-free formation of a millimeter-scale multicellular spheroid with an internal cavity from magnetically levitated 3T3 cells that ingested iron oxide-containing microspheres. *Biotechnol. Bioeng.* **111**, 1038–47 (2014).
40. Guo, W. M., Loh, X. J., Tan, E. Y., Loo, J. S. C., and Ho, V. H. B. Development of a magnetic 3D spheroid platform with potential application for high-throughput drug screening. *Mol. Pharm.* **11**, 2182–9 (2014).

41. Yarrow, J. C., Perlman, Z. E., Westwood, N. J., and Mitchison, T. J. A high-throughput cell migration assay using scratch wound healing, a comparison of image-based readout methods. *BMC Biotechnol.* **4**, 21 (2004).
42. Soderholm, J. and Heald, R. Scratch n' screen for inhibitors of cell migration. *Chem. Biol.* **12**, 263–5 (2005).
43. Huang, C., Rajfur, Z., Borchers, C., Schaller, M. D., and Jacobson, K. JNK phosphorylates paxillin and regulates cell migration. *Nature* **424**, 219–223 (2003).
44. Lampugnani, M. G. Cell migration into a wounded area *in vitro*. *Methods Mol. Biol.* **96**, 177–82 (1999).
45. Hausman, D. B., DiGirolamo, M., Bartness, T. J., Hausman, G. J., and Martin, R.J. The biology of white adipocyte proliferation. *Obes. Rev.* **2**, 239–54 (2001).
46. Zuk, P. A. et al. Multilineage cells from human adipose tissue: Implications for cell-based therapies. *Tissue Eng.* **7**, 211–28 (2001).
47. Daquinag, A. C., Zhang, Y., Amaya-Manzanares, F., Simmons, P. J., and Kolonin, M. G. An isoform of decorin is a resistin receptor on the surface of adipose progenitor cells. *Cell Stem Cell* **9**, 74–86 (2011).
48. Traktuev, D. O. et al. A population of multipotent CD34-positive adipose stromal cells share pericyte and mesenchymal surface markers, reside in a periendothelial location, and stabilize endothelial networks. *Circ. Res.* **102**, 77–85 (2008).
49. Menscher, A. L. *Junqueira's Basic Histology: Text and Atlas.* New York, NY: McGraw-Hill Lange, 2009, 298–315.
50. Butcher, J. T. and Nerem, R. M. Valvular endothelial cells regulate the phenotype of interstitial cells in co-culture: Effects of steady shear stress. *Tissue Eng.* **12**, 905–15 (2006).
51. Flanagan, T. C. et al. A collagen-glycosaminoglycan co-culture model for heart valve tissue engineering applications. *Biomaterials* **27**, 2233–46 (2006).
52. Frearson, J. A. and Collie, I. T. HTS and hit finding in academia–from chemical genomics to drug discovery. *Drug Discov. Today* **14**, 1150–8 (2009).

Index

A

Aberrant mechanotransduction
  processes, 208
Absorbed energy, 51
Acrylamide, 156
Action potentials, 54–55
Actomyosin, 61
  blocker, 82
Adaptable control method, 212
Adhesion molecules, 61
Adipose tissue, 238
Airborne magnetic dusts, 62
Alkyne, 25
Alloy steels, 137
Alternating frequency (AC) magnetic
  field, 1
Amidation, 24
Amide bonds, 24
Amines, 19, 24, 25
Ammonium group, 20
Ammonium persulfate
  (AMPS), 156
Amoeboid chemotaxis, 209
AMPS, *see* Ammonium persulfate
Anchoring groups, 19–20
Angiogenesis, 149
Angioplasty, 135
Angular deformation, 68
Anti-β1 integrin antibody-coated
  magnetic beads, 209
Antibodies, 148
Antibody-conjugated
  magnetoliposomes, 215
Antiferromagnetism, 3
Aortic valve, 239
Aqueous coprecipitation, 13
Aqueous-dispersed nanoparticles,
  158–160
Arginine-glycine-aspartic acid (RGD)-
  containing peptide-coated
  beads, 209
Autologous cells, 124
Azide, 25

B

Bead, 66–67
  twisting, 79, 85, 86
Beer's law, 157
Benzyl ether, 155
Biocompatible polymers, 2
Bioengineering applications, cellular
  patterning for, *see* Magnetic
  cell patterning
Biological media, 19
Biomedicine
  magnetism and, 1–4
  micro- and nanoscale magnetic
    bio-actuators, 4–7
Biomolecule-functionalized magnetic
  nanoparticles, 205
Biotin acceptor peptide, 49
Bitter technique, 175
Blood proteins, 159
Bohr magnetons, 3, 12
Boltzmann's constant, 178
Bond dissociation, 74
Bone, formation and remodeling, 208
Bottom-up approaches, 10, 105–106
  electrodeposition, 106–107
  physical vapor deposition, 107
  powder metallurgical and casting
    processes, 106
Bovine serum albumin (BSA), 182
Bronchiole, 238–239
  coculture model, 215, 235, 238
Brownian motion, 22, 177, 181
Brownian relaxation, 46
BSA, *see* Bovine serum albumin

C

Caenorhabditis elegans, 55
Calcific aortic valve disease (CAVD), 239
Calcium, 43, 55
Calcium channels, 43
Cancer therapies, 149
Carboxylated ferromagnetic beads, 63

Carboxylates, 19–20
Carboxylic acid, 24
Casting process, 106
Catechols, 19, 20
Cations, 11
CAVD, *see* Calcific aortic valve disease
Cell-based therapy, 123
   applications, 135
   cell loading, 136
   magnetically responsive materials, 124–127
   magnetic cell delivery, 127
   magnet system, 136–139
   MNPs and therapeutic cells, requirements, 131–133
   potential, 124
Cell–cell interactions, 181
Cell-encapsulation hydrogels, 215
Cell loading, 136
Cell magnetometry, 62
Cell–matrix adhesion, 61
Cell–matrix interactions, 83
Cell membrane, 53
   targeting mechanoreceptors, 208
Cell-seeded biomaterial scaffolds, 206
Cell signalling, via ion channels, 42–43
Cell-surface receptors, 208
Cell twister, 66–67
Cellular calcium influx, remote triggering, 53–54
Cellular deformation, 71
Centrifugation, 180
Cetyltrimethylammonium bromide (CTAB), 17
Chemical etching, 108
Chemomechanical kinetic model, 209
Chemotherapy, 148, 151
Chitosan, 18
Citrate-coated Fe_2O_3 nanoparticles, 181
Click-chemistry approach, 25, 26, 49
Clustering, 84
Cobalt(II) acetylacetonate, 155
Cobalt ferrite NPs, 155
Co grains, 198
Colloidal magnetic fluids, 178, 192
Colloidal state, characterization of, 22–23
Colloidal systems, 70
Comprehensive cellular model, 60

COMSOL Multiphysics, 96
Conduction pore, 42
Conjugation techniques, 24
Constant temperature drug release, 157, 165
Contrast matching, 23
Conventional thermocouple, 51
Copper catalyst, 25
Copper-free reactions, 25
Core–shell geometry, 48
Correlation function, 22
Crosslinked PNIPAAm hydrogels, 152
CSK, *see* Cytoskeleton
CTAB, *see* Cetyltrimethylammonium bromide
Cytochalasin D, 62, 69, 70
Cytoskeleton (CSK), 60, 61, 66, 79

D

DAB, *see* Diaminobenzidine
Debye force, 21
Deionized (DI) water, 156
Dendritic cells, 123
Dendritic processes, 208
Derjaguin–Landau–Verwey–Overbeek (DLVO) theory, 21
Dextran, 18
Diamagnetic materials, 125
Diaminobenzidine (DAB), 232
Diffusion-limited aggregation, 182
2,3-Dimercaptosuccinic acid (DMSA), 18, 155
Dipole interactions, 21
Disk drive, 198
DLS, *see* Dynamic light scattering
DMSA, *see* 2,3-Dimercaptosuccinic acid
DNA bioconjugation, 25
DNA-damaging agent, 184
Domain walls (DWs), 125, 126
Double-Voigt model, 68, 69, 72
Drug delivery vehicles, 124
Drug-eluting stents, 135
Drug loading, thermosensitive drug carriers, 157
Drug release, constant temperature and pulsatile, 157, 165–166
DWs, *see* Domain walls
Dynamic force spectroscopy, 72

Dynamic light scattering (DLS), 21–23, 49, 154, 160
Dynamic rheology model, 60

E

ECM, *see* Extracellular matrix
Electrical discharge machining (EDM), 108
Electrodeposition, 106–107, 194
Electromagnetic needles
    advantages, 101–102
    field production, 99
    limitations, 102
    magnetization curve, 100
    simulation, 101
Electromagnetic switch, 212
Electromagnets, 128, 186
Electronics, 193
Electrons, 125
Electrostatic interactions, 21
ELF, *see* Extremely low frequency
EMT, *see* Endothelial-mesenchymal transdifferentiation
Endocytosis, 66
Endothelial cell(s), 124, 133, 184
    adhesion, 185
Endothelial integrity, 135
Endothelial-mesenchymal transdifferentiation (EMT), 239
Endothelial progenitor cells (EPCs), 212
Enhanced permeation and retention (EPR) effect, 149
Environmentally responsive polymers, 152
EPCs, *see* Endothelial progenitor cells
Epidermal growth factor (EGF)-coated magnetic beads, 209
Epithelial cells (EpiC), 238
Equal-sized magnetic beads, 191
Equilibrium polymer fractions, 157
Equilibrium swelling, 156–157, 163–165
Erythrocytes, 124
Ethanol, 155
1-Ethyl-3- (3-dimethylaminopropyl) carbodiimide (EDC), 24
Eukaryotic cells, 60

Experimental magnetic capture, thermosensitive drug carriers, 153–154, 159
External permanent magnets, 128
Extracellular matrix (ECM), 230
    deposition, 215
Extremely low frequency (ELF), 216

F

Far from equilibrium, 78
FBS, *see* Fetal bovine serum
FEA, *see* Finite-element analysis
Femtosecond laser, 184
Ferrimagnetic materials, 125
Ferrimagnetism, 3
Ferrite materials, 48
Ferrite nanoparticles
    additional modification, 24–25
    anchoring groups, 19–20
    aqueous coprecipitation, 13
    colloidal state, characterization of, 22–23
    ferrite crystal structure, 11–12
    inorganic modification, 17–18
    interparticle interactions, 21–22
    organic modification, 18–19
    surface modification, 16–17
    thermal decomposition, 13–15
Ferrites, 10, 112
Ferrofluids, 10, 189, 190
Ferromagnetic materials, 2, 112, 125
Ferromagnetic microbeads, 63, 68
Ferromagnetic wires, 177
Ferromagnetism, 3
Fetal bovine serum (FBS), 153, 160, 161
Filtration technique, 177
Finite-element analysis (FEA), 96
Finite-element simulations, 94
Flow cytometry, 229
FluidMAG-D®, 155, 160
Fluorescence, 51
    quantum yield, 51–53
Fluorophores, 51
5-Fluorouracil (5-FU), 157, 166
Folate receptors, 151
Free radical solution polymerization, 156
5-FU, *see* 5-Fluorouracil
Furlani group, 188, 194

G

Gadolinium (Gd) nanoparticles, 229
Gas-phase synthesis methods, 10
Gaussmeter, 153
GBM, *see* Glioblastoma
GFP-transfected human GBM, 236
Glioblastoma (GBM), 231, 236–238
Gold, 18
Green fluorescent protein (GFP)
    fluorescence intensity, 53

H

Halbach array, 130, 136
Hard ferromagnets, 95
Heating-up method, 14
Heat transfer, 47–48
HEK293 cells, 54
Helmholtz coils, 63–65, 85
H-field (magnetizing field), 100, 101
HGMS, *see* High-gradient magnetic
    separators
High-frequency magnetic twisting
    cytometry, 59
High-gradient magnetic separators
    (HGMS), 177
Highly efficient fibroblast, 215
Highly phagocytic cells, 133
High-resolution transmission electron
    microscopy (HRTEM), 16
High-throughput toxicity testing, 239
hMSCs, *see* Human mesenchymal stem
    cells
Horizontal Helmholtz coils, 63–65, 85
Hormone, remote control, 55, 56
Hot-injection reactions, 14
HPMECs, *see* Human pulmonary
    microvascular endothelial cells
Human airway smooth muscle (HASM)
    cells, 85
Human diseases, 42
Human mesenchymal stem cells
    (hMSCs), 133, 218
Human pulmonary microvascular
    endothelial cells (HPMECs), 67,
    78–81
Hydrodynamic model, 62
Hydrodynamic radius effect, 48

Hydrogels, in magnetic nanoparticles,
    156, 163
2-Hydroxyethyl methacrylate, 156
Hyperthermia, 152, *see also* Magnetic
    NP-based hyperthermia

I

ICP-MS, *see* Inductively coupled plasma
    mass spectrometry
Immune tolerance, 124
Immunohistochemical staining, 232
Induction welding, 51
Inductively coupled plasma mass
    spectrometry (ICP-MS), 49
Inorganic coatings, 17
Inorganic modification, 17–18
Instantaneous magnetic field, 66
In-stent stenosis, 138
Insulin, gene expression, 55, 56
Integrins, 82, 84, 212
Intensity-based scattering, 23
Interleaved bilayer templated assembly,
    197
Interleukin-1β (IL-1B), 218
Intermolecular chemical potential, 75
Interparticle interactions, 21–22
Intimal hyperplasia, 135
Intimate covalent bonds, 72
Intracellular C-terminus tail region,
    209
Intracellular magnetic particles, 62
In vivo magnetic cell stimulation, 42,
    57
In vivo rabbit model, 212
Ion channel
    cell signaling via, 42–43
    stimulation, 44
Iron acetyl acetonates, 14, 155
Iron-containing nanoparticles, 152
Iron oleate, 15
Iron oxide-based nanoparticles, 229
Iron oxide magnetic nanoparticles,
    212
Iron oxides, 4
Iron pentacarbonyls, 14
Iron pin, 182
Iron platinum NPs, 155
N-Isopropylacrylamide, 152, 156

K

Keesom force, 21
Kimwipe®, 157
Kramer's theory, 75

L

LaMer model, 13, 14
Laser machining, 99, 107–108
LCST, *see* Lower critical solution
        temperature
Ligand binding, 84
Ligand-exchange reaction, 155
Ligand-targeted nanoparticle systems,
        149
Light scattering, 22
Liquid dielectrics, 108
Lithographed microtoroid, 186
Lithographic techniques, 10
Localized hyperthermia, 151
Longitudinal media patterns, 198
Longitudinal recording, 198
Lower critical solution temperature
        (LCST), 151, 152, 163
Low-molecular-weight stabilizers, 20

M

Macroscopic permanent magnets, 193
MACS, *see* Magnetically assisted cell
        sorting
Maghemite, 159
    nanoparticles, 127, 131
    nanopowder, 153, 154, 161
Magnet arrays, 130, 216
Magnetic activation of receptor
        signaling (MARS), 5, 207, 216
Magnetic actuation, 151
Magnetically assisted cell sorting
        (MACS), 229
Magnetically blocked particles, 211
Magnetically responsive materials,
        124–131
Magnetic beads, 61, 63, 184
Magnetic bioreactor, 221
Magnetic capture, 151, 169
    thermosensitive drug carriers, 153,
        158–160

Magnetic cell culture, 231–235
Magnetic cell delivery, 127
Magnetic cell patterning
    applications, 198
    micrometer-scale pattern, 193
        colloidal magnetic fluids, 192
        equal-sized magnetic beads, 191
        ferrofluids, 189, 190
        Furlani group, 188, 194
        rod magnets, 187
        toroids, 186
    millimeter-scale pattern
        endothelial cell spheroids, 178–179
        magnetic tips, 181
        magnetite cationic liposomes, 182
        negative magnetophoresis,
            182–183
        polydimethylsiloxane, 184
        position magnets, 179
        RGD adhesion, 185
        stem cell spheroids, 179–180
    nanometer scale magnetic-field-
        directed self-assembly, 193–198
Magnetic dipole
    magnetic field, 117
    yields
        loop model for, 113
        magnetic charge model for, 114
Magnetic disk, 195
Magnetic energy transfer, 211
Magnetic field-directed self-assembly,
        177, 178
Magnetic field gradient assembly, 177
Magnetic fluid hyperthermia (MFH), 2,
        148, 211
Magnetic flux, 109–111
Magnetic force, 113–115, 176, 231, 235
Magnetic force bioreactor (MFB), 216,
        218
Magnetic force-induced cell levitation,
        215
Magnetic heating, 158, 167–168
Magnetic inhomogeneities, 175
Magnetic interactions, 21
Magnetic ion channel activation
        (MICA), 5, 207, 216, 217
Magnetic levitation, 6, 232–235
Magnetic localization, 150
Magnetic microbead, 183, 185

Magnetic microwells, 188
Magnetic moment, 126
Magnetic nanoparticle (MNP)-loaded
    cells, 128
Magnetic nanoparticles (MNPs), 5, 6,
    42, 198
  accumulation of, 150
  applications, 9
  biocompatible surface coating, 222
  cancer research, 148
  cell-based therapy, 131–133
  chains of, 194
  dispersion, 155
  ferrites, 10
  in hydrogels, 156, 163
  magnetic energy transfer, 211
  MARS/MICA, 216, 220, 221
  synthesis of, 17
  targeting mechanoreceptors, 208
  in tissue engineering and
      regenerative medicine,
      211–216
  trapping, 186
Magnetic nanowires, 215
Magnetic needle, 94
  applications and calculations, 113
    magnetic forces, 113–115
    magnetic particle capture, 115–120
  bottom-up approaches, 105–106
    electrodeposition, 106–107
    physical vapor deposition, 107
    powder metallurgical and casting
      processes, 106
  electromagnets, 100
    advantages, 101–102
    field production, 99
    limitations, 102
    magnetization curve, 100
    simulation, 101
  geometry, 109–112
  material, 112–113
  permanent magnets, 95–96
    advantages, 98
    limitations, 98–99
    magnetic portion, 95
    simulation, 96–98
  soft-magnetic needles, 102–103
    advantages, 104
    limitations, 104–105

    potential configurations, 103
    simulation, 103–104
  top-down approaches, 107
    chemical etching, 108
    laser machining, 107–108
Magnetic NP-based hyperthermia,
    44–45
  heat transfer, 47–48
  power dissipation, 45–46
  relaxation mechanisms, 46–47
Magnetic particle capture, 115–120
Magnetic resonance imaging (MRI), 1,
    2, 212
Magnetic stimulation, 42
Magnetic susceptibility, 114
Magnetic 3D bioprinting, 232, 233, 239
Magnetic tips, 181
Magnetic torque, 66
Magnetic twisting cytometry (MTC), 5,
    59–61
  bead, as cell twister and transducer,
    66–67
  experimental device, 63–66
  features of, 83–85
  historical aspects, 61–63
  living cell response, characterization
    of, 79–83
  mechanical and adhesion properties,
    71–72
  microrheological viscoelastic models,
    68–71
  molecular adhesion models, *see*
    Molecular adhesion models
Magnetism, 1–4, 192
Magnetite cationic liposomes (MCLs),
    182
Magnetite nanoparticles, 127, 131
Magnetizable implants, 130, 137
Magnetizable stent, 131
Magnetomechanical stimulation, 207
Magnetometry, 21
Magnetopneunography, 62
Magnetothermal-triggered delivery,
    169
Magnet system, 127, 136–139
Magnitude
  color shades and, 101, 103
  of induction, 45
  narrow magnet, 129

Manufacturing techniques, magnetic
        needles, 94
    bottom-up approaches, 105–106
        electrodeposition, 106–107
        physical vapor deposition, 107
        powder metallurgical and casting
            processes, 106
    top-down approaches, 107
        chemical etching, 108
        laser machining, 107–108
MARS, *see* Magnetic activation of
        receptor signaling
Matrigel™, 215
MBAAm, *see* Methylene bisacrylamide
mCherry-transfected NHAs, 236
MCLs, *see* Magnetite cationic liposomes
Mechanical energy, 75
Mechanical stimulation, 82, 206–208
Mechanoactivation technology, 222
Mechanoresponsive proteins, 209
Mechanosensitive channel large
        conductance (MscL), 216
Mechanotransduction, 61, 208
Membrane hyperpolarization, 218
Membrane potential, 43
Membrane voltage, 43
Mercaptoundecanoic acid, 155
Mesenchymal stem cells (MSCs), 206
Metal salt reduction, 10
Methylene bisacrylamide (MBAAm),
        156
MFB, *see* Magnetic force bioreactor
MFH, *see* Magnetic fluid hyperthermia
M-field, 100, 102
M-gels, 215
MICA, *see* Magnetic ion channel
        activation
Micro-magnetic bio-actuators, 4–7
Micromanipulation techniques, 59
Microparticles, 42
Micropatterned magnets, 189
Microrheological viscoelastic models,
        68–71
Microscale, patterning magnets at, 193
Mimosine, 20
Mitogen-activated protein kinase, 218
Mitomycin C (MMC), 184
MLCK, *see* Myosin light chain kinase
MMC, *see* Mitomycin C

MNP-loaded cells, 130, 131
MNPs, *see* Magnetic nanoparticles
Mobile device-based imaging system, 240
Molecular adhesion models, 72
    effect of force, 73, 75–76
    experimental conditions, 77
    ramp rate, 78–79
    sliding effect, 74
    twisting stress, time course of, 78
Molecular dynamics, 63
Molecular motors, 61
Monochromatic laser, 22
MSCs, *see* Mesenchymal stem cells
MTC, *see* Magnetic twisting cytometry
Multicomponent self-assembly, 190
Murine embryonic stem cells, 180
Myosin II motors, 84
Myosin light chain kinase (MLCK), 82

N

Nanometer scale magnetic-field-
        directed self-assembly, 193–198
Nanoparticle (NP), 209
    heating capacity, 50–51
    physical and magnetic
        characterization, 49–50
    synthesis and functionalization, 48–49
Nanoparticle-based heating, 43–44
Nanoscale axial displacement, 60
Nanoscale field gradients, 177
Nanoscale magnetic bio-actuators, 4–7
Nanoscale magnetic particles, 4
Nanoscale templated self-assembly, 196
NdFeB permanent magnets, 128, 129
Neél-Arrhenius relationship, 3
Neel relaxation, 46, 152
Negative magnetophoresis, 182, 183
Neurons, 43
Newtonian particle transport, 178
NHAs, *see* Normal human astrocytes
Nickel, 176
Nickel-coated stainless steel stents, 137
Nickel ferrite, 12
Nitrocatechol, 20
Nitrone-based materials, 25
Nonmagnetic beads, 191
Normal human astrocytes (NHAs), 235,
        236, 238

O

Oleic acid, 18
Oleyl amine, 155
Opsonization, of magnetic
  nanoparticles, 154–155, 160–162
Optical diffraction, 196
Optical methods, 42
Optics, 193
Optimal heating, 48
Organic coatings, 17
Organic modification, 18–19
Organic solvents, 18
Osteocyte, 208
Osteogenic media, 220
Ostwald ripening, 15
Oxygen ions, 13

P

Paramagnetic materials, 125
Paramagnetic particles, 191
Paramagnetism, 3
Pattern transfer nanomanufacturing™
  (PTNM), 196
PBS, *see* Phosphate buffered saline
PDMS, *see* Polydimethylsiloxane
PECs, *see* Pulmonary endothelial cells
PEG, *see* Polyethylene glycol
PEI, *see* Polyethylenimine
Permanent deformation, 71
Permanent magnet needles, 95–96
  advantages, 98
  limitations, 98–99
  magnetic portion, 95
  simulation, 96–98
Permanent magnets, 150
Perpendicular recording, 198
PFs, *see* Pulmonary fibroblasts
PHEMA, *see* Polymer gels of
  2-hydroxyethyl methacrylate
Phosphate buffered saline (PBS), 20
Phosphates, 20
Phosphonates, 19
Photo bleaching, 53
Photon correlation spectroscopy (PCS),
  see Dynamic light scattering
Physical vapor deposition (PVD)
  processes, 107

PLL, *see* Poly-L-lysine
Poly(NIPAAm-co-acrylamide), 152, 156,
  162
Polyacrylamide, 152
Polyak group, 137
Polydimethylsiloxane (PDMS), 180
Polyethylene glycol (PEG), 18, 49, 149,
  153, 159
Polyethylene oxide (PEO), 18, 20
Polyethylenimine (PEI), 20
Poly-L-lysine (PLL), 231
Polymer, 18, 20
  film, 196
  scaffolds, 231
  systems, 151
Polymer-based magnetic nanoparticles,
  131
Polymer gels of 2-hydroxyethyl
  methacrylate (PHEMA), 156,
  158, 162
Polystyrene microparticles, 180
Porcine endothelial cells, 215
Powder metallurgical process, 106
Power dissipation, magnetic NP-based
  hyperthermia, 45–46
Power law model, 69, 70, 72, 82
Protein gels, 231
Protein production, 55–56
Protoplasm, 5
Pulmonary endothelial cells (PECs), 238
Pulmonary fibroblasts (PFs), 238
Pulsatile drug release, 157, 165
PVD, *see* Physical vapor deposition
Pyridyl disulfide, 25

Q

Quasi-elastic light scattering (QELS), *see*
  Dynamic light scattering

R

Radio-frequency (RF) magnetic field, 2,
  42, 43, 54, 56, 211
Rare-earth alloys, 113
Rare-earth magnets, 99, 128
Rare-earth permanent-magnetic
  materials, 104
Recording medium, 195

Relaxation mechanism, 46–47, 62
Remanent magnetic field, 66
Resonance, 1
Restenosis, 135
Reticuloendothelial system (RES), 16
RGD cell-binding peptide, 184
RGD-functionalized superparamagnetic
     microbeads, 215
RGD ligands, 79
Ring closure, 240
Rod magnets, 187

S

SANS, *see* Small angle neutron
     scattering
SAR, *see* Specific absorption rate
Saturation magnetization, 49
SAXS, *see* Small angle x-ray scattering
Scattered light, 22
Self-assembly, 176
SGR model, *see* Soft glass rheology
     model
Silanes, 19
Silica, 17
Single-dimensional velocity, 116
Single-Voigt model, 68, 69
Sliding angle, 77
Sliding effect, 74
Small angle neutron scattering (SANS),
     23
Small angle x-ray scattering (SAXS), 23
Smooth muscle cells (SMCs), 67, 135, 231,
     239
Soft glass rheology (SGR) model, 60
Soft-magnetic needle, 102–103
     advantages, 104
     limitations, 104–105
     potential configurations, 103
     simulation, 103–104
Soft magnets, 95
Solenoid coils, 51
Specific absorption rate (SAR), 48, 50
Spherical ferromagnetic particles, 63
Spheroid aggregation, 180
Spheroid systems, 231
Spinel unit cell, 11
Stainless steel stents, 137
Static elasticity theory, 67

Static magnetic fields, 1, 150
Static neodymium magnets, 153
Stem-cell differentiation, 212
Stem cell spheroids, 180
Stemness, 133
Steric interactions, 21
Stokes–Einstein relation, 22
Streptavidin-biotin linkage, 44
Stromal vascular fraction (SVF), 238
Subcellular biology, magnetic field
     gradients for, 186
Superexchange mechanism, 12
Superparamagnetic carrier particles, 127
Superparamagnetic iron oxide (SPIO)
     microspheres, 215
     nanoparticles, 133
Superparamagnetic microbeads, 188
Superparamagnetic nanoparticles, 3, 4,
     49, 127
     hyperthermia, 44
Surface modification, 16–17
SVF, *see* Stromal vascular fraction

T

Targeting magnet system, 128
Targeting mechanisms
Targeting mechanoreceptors, 208–211
Teflon®, 156
TEM, *see* Transmission electron
     microscopy
TEMED, *see* N,N,N',N'-
     Tetramethylethylenediamine
Temperature dependence, 51, 52
Temperature-sensitive ion channel,
     43–44
Tendinopathy, *in vivo* rabbit model, 212
Tensegrity approach, 85
Tensile forces, 208
TERM, *see* Tissue engineering and
     regenerative medicine
Tetraethyl orthosilicate (TEOS), 17
N,N,N',N'-Tetramethylethylenediamine
     (TEMED), 156
Theophylline, 157, 166
Therapeutic cells, 127
     cell-based therapy, 131–133
Thermal decomposition, 10, 13–15, 27,
     155

Thermomagnetic activation
  cell signaling via ion channels,
    42–43
  magnetic NP-based hyperthermia,
    44–45
    heat transfer, 47–48
    power dissipation, 45–46
    relaxation mechanisms, 46–47
  remote triggering
    action potentials and behavior,
     54–55
    cellular calcium influx, 53–54
    protein production and secretion,
     55–56
  thermomagnetic stimulation, *see*
    Thermomagnetic stimulation
Thermomagnetic stimulation, 43–44
  local heating measurement, 51–53
  nanoparticles
    physical and magnetic
     characterization, 49–50
    surface, local temperature, 53
    synthesis and functionalization,
     48–49
  specific absorption rate, 50–51
Thermometer, 53
Thermosensitive anion channels, 43
Thermosensitive drug carriers,
  148–151
  constant temperature and pulsatile
    drug release, 157, 165
  equilibrium swelling, 156–157,
    163–165
  hydrogel synthesis, 156, 162
  magnetic capture
    aqueous-dispersed nanoparticles,
     158–160
    experimental, 153–154
  magnetic heating, 158, 167–168
  magnetic nanoparticles
    opsonization, 154–155
    synthesis, 155, 162
  protein opsonization effect, 160–162
  targeting mechanism, 148–151
  triggering mechanism, 151–153
Thermosensitive protein, 47–48
Thiol, 19, 25
Thiol-based conjugation, 27

Three-dimensional (3D) cell culture,
  230, 231, 240
  adipose tissue, 238
  aortic valve, 239
  bronchiole, 238–239
  glioblastoma, 236–238
  magnetic cell culture, 231–235
  wound healing assay, 239–240
Three-dimensional (3D) endothelial cell
  spheroids, 178, 179
Time-of-flight formula, 117, 119
Time-varying external magnetic fields,
  218
Tissue culture, 179
Tissue engineering and regenerative
  medicine (TERM), 206–207,
  222
  magnetic nanoparticles in,
    211–216
Tissue healing, 212
Top-down approaches, 107
  chemical etching, 108
  laser machining, 107–108
  magnetic nanoparticles, 10
Tracheal smooth muscle cells, 238
Transducer, 66–67
Transition-metal alloys, 112
Translational force, 218
Transmembrane adhesion receptors,
  79
Transmission electron microscopy
  (TEM), 16, 49
Trialkoxysilylpropanes, 17
Triggering mechanisms, 151–153
Trivalent cations, 11
TRP channel, 43
TRPV1 channel, 43, 44, 54
TWIK-Related K$^+$ Channel 1 (TREK1),
  209, 210
Twisting field, 65
Two-component ring assembly tuning,
  191
Tygon tubing, 153
Type II USP dissolution system, 157

U

UV/Vis spectrophotometry, 156

V

Valve interstitial cells (VICs), 239
Van der Waals forces, 21
Vascular endothelial cells (VECs), 239
Vascular network formation, 215
VECs, *see* Vascular endothelial cells
VICs, *see* Valve interstitial cells
Viscoelastic models, 68
Viscoelastic solid-like system, 66, 72, 73
Voltage-gated channels, 43

W

WAT, *see* White adipose tissue

Water-cooled coil, 51
Wet-chemistry methods, 10, 13
White adipose tissue (WAT), 238
Wnt signal transduction pathways, 210
Wound healing assay, 239–240

Y

Young's modulus, 67

Z

Zeta potential measurements, 21
Zipper bond, 76

Printed and bound by CPI Group (UK) Ltd, Croydon, CR0 4YY

01/11/2024

01782617-0004